# DATA & CHANCE

with manipulatives

Beth McCord Kobett
Kay B. Sammons

Published by Ideal School Supply
an imprint of

Beth McCord Kobett is a former mathematics Resource Teacher for the Howard County Public School System in Ellicott City, Maryland. As a resource teacher, she conducted individual performance assessments, demonstration lessons, parent meetings, and school inservices for thirty elementary schools. She has taught fourth and fifth grade. She currently conducts workshops in the use of math manipulatives, problem solving, collecting and organizing data, probability, and the use of literature and writing using graphic organizers in the mathematics classroom. She also teaches methods for elementary mathematics at Johns Hopkins University School of Continuing Studies.

Beth holds a bachelor of science degree from the University of Missouri and a master's degree from Johns Hopkins University.

Kay B. Sammons is currently the elementary mathematics supervisor for the Howard County Public School System in Ellicott City, Maryland. As a supervisor, she is responsible for curriculum development and staff development for elementary school teachers. Previously, Kay was the elementary mathematics resource teacher. She has also taught fourth, fifth, and sixth grade. Kay conducts workshops on performance assessment and using manipulatives to teach mathematics, estimation, and number sense. She is an itinerant instructor for Johns Hopkins University.

Kay holds a bachelor of science degree from the University of Maryland. She also holds a master's degree from the University of Maryland and a master's degree from Johns Hopkins University.

---

*McGraw-Hill*
***Children's Publishing***
*A Division of The* ***McGraw-Hill*** *Companies*

Published by Ideal School Supply
An Imprint of McGraw-Hill Children's Publishing


Design Supervisor: Annelise Palouda
Illustrations: Donna Reynolds and Mark Keli'ihanapule
Production Coordinator: Alison Tudury
Editor: Susan Blackaby
Project Manager: Shirley Hoogeboom
Art Director: Nancy Tseng



Send all inquiries to:
McGraw-Hill Children's Publishing
3195 Wilson Drive NW
Grand Rapids, MI 49544



*Math Discoveries About Data and Chance*—grades 3-4
ISBN: 1-56451-181-2

# CONTENTS

# NOTES TO THE TEACHER

*Math Discoveries About Data and Chance, Grades 3-4,* is part of a series of Math Discoveries books. Each book focuses on one area of the mathematics curriculum. They are designed to help your students build mathematical concepts and understandings through hands-on activities with concrete models and tools. The activities in these books engage students in "doing mathematics," a learning method that the National Council of Teachers of Mathematics (NCTM) emphasizes in its Curriculum and Evaluation Standards for School Mathematics. The activities also emphasize problem solving, communication, reasoning, and making mathematical connections.

Using models and tools invites the students to explore, solve problems, construct, discuss, investigate, describe, represent, and predict. The activities are designed for students working together in pairs and small groups, which encourages students to share their thinking and learning. By manipulating models and tools, using their own language to explain their thinking, and sharing their thinking and learning with others, students build deeper mathematical understandings and develop communication skills. The concrete materials serve as a focus for communication, even among the students who do not share a primary language.

Each book presents 40 reproducible one-page explorations for students, three of four investigation for more in-depth explorations, teaching notes, and blackline masters for special recording sheets. Sample problem solutions are also included.

## INTRODUCTION TO *MATH DISCOVERIES ABOUT DATA AND CHANCE, GRADES 3–4*

The explorations and investigations in this book are designed to help students develop the math skills and understanding articulated in the NCTM standards.

**NCTM STANDARDS**

| NCTM STANDARDS FOR MATH SKILLS AND UNDERSTANDINGS | ACTIVITIES THAT DEVELOP THESE SKILLS AND UNDERSTANDINGS |
|---|---|
| • Determine outcomes and chance | • Explorations 1-5, Investigation 3 |
| • Predict outcomes | • Explorations 6-10, Investigation 3 |
| • Determine fairness | • Explorations 11-15, Investigation 3 |
| • Interpret and construct pictographs, line plots | • Explorations 16-20, Investigation 2 |
| • Interpret and construct glyphs, bar graphs | • Explorations 21-25, Investigation 1 |
| • Interpret and construct circle graphs | • Explorations 26-30 |
| • Determine measures of central tendency | • Explorations 31-35, Investigation 2 |
| • Compare and apply graphs | • Explorations 36-40, Investigation 1, 4 |
| • Relate math to their daily lives | • Investigation 1, 2, and 4 |
| • Collect, organize, and display data | • Investigation 1, 2, and 4 |

Manipulatives are wonderful tools and models for students to use. They enable students to explore concepts in ways that would be impossible on paper. When students are able to hold something in their hands, to look at it and move it around in various ways, and to take it apart, their learning is greatly enhanced. They are able to investigate and explore concepts in concrete ways; this allows them to integrate concepts into their basic math understandings from a sound base of personal experience. The intrinsic value of manipulatives in math education is the underlying premise of this book.

The students will be using dice to determine possible outcomes and test predictions. Transparent spinners can be placed over spinners found on exploration pages or on student-made spinners. LacerLinks are used to develop cube graphs that lead into bar graphs and circle graphs. LacerLinks will be placed in a paper bag and drawn out at random to determine outcomes. These materials are integral to developing the concepts of probability and graphing. Students need to use a variety of materials to demonstrate a transfer of the concepts of probability. In addition, the materials are used to develop various graphs at the concrete level. Students will also need crayons or markers, paper bags, and blank paper to draw on for making spinners.

USING *MATH DISCOVERIES ABOUT DATA AND CHANCE, GRADES 3–4*

**Contents**

This book contains 40 short problem-solving activities—explorations, divided into eight sections. There are five similar explorations in a section. Each section of explorations is preceded by teaching notes that identify the skills and understandings developed in the explorations and describe an activity for introducing the explorations. The teaching notes may also suggest questions that can be used to encourage students to think about and communicate about what they have learned, provide ideas for extending the activities, and suggest ways to assess student learning.

This book also contains four longer, more open-ended activities—investigations. The investigations give students opportunities to extend and deepen their learning and to apply what they have learned to solving a problem.

**Suggestions for Classroom Use**

These activities can be used by students working individually, in pairs, or in small cooperative learning groups. Working together encourages students to talk about their thinking and about their discoveries. Students will benefit from articulating their thinking and hearing how others may have solved the same problem in a different way. To take advantage of this, encourage the students to share their ideas with other pairs of students, with other small groups, or with the whole class.

Open-ended questions that invite students to express their thoughts are critical to the classroom discussion. Frequently the activities have a single solution, but there are a multitude of solution paths. As students share the insights they have gained through developing their solutions, they teach one another important concepts about mathematics.

If the students have not used particular manipulatives before, give them time to become familiar with them. Then let students begin the explorations.

## Materials Needed

It is recommended that each student or pair of students have:

dice
transparent spinners
LacerLinks
crayons or markers
paper bags

You can make copies of the explorations for each student or pair of students to use, or you can place copies in a learning center. Students may record directly on the front and back of the exploration pages.

For Investigation 1, students will need:

bags of popcorn ( 3 brands of pre-popped or microwave)
popcorn glyph key
crayons or markers
*The Popcorn Book* by Tomie de Paola

For Investigation 2, students will need:

calculators
*Averages* by Jane Jonas Srivastava

For Investigation 3, students will need:

a variety of board games
cardboard
construction paper
glue
scissors
crayons or markers
dice, transparent spinners, cards

For Investigation 4, students will need:

graph paper
crayons or markers

 

# SAMPLE SOLUTIONS

## EXPLORATIONS 1-5

EXPLORATION 1:

A. 1,1,1,1
   1 out of 4
   1 out of 4
   1 out of 4
   1 out of 4

B. No. Each color has an equally likely chance.

C. Answers will vary.

EXPLORATION 2:

A. 1 chance out of 4
   1 chance out of 4
   1 chance out of 4
   1 chance out of 4

B. green, green, blue, purple

C. red, red, blue, purple, purple

EXPLORATION 3:

A. Answers will vary.

B. 1 out of 6
   1 out of 6
   1 out of 6
   1 out of 6
   1 out of 6
   1 out of 6
   3 out of 6
   3 out of 6

C. No. The chances of rolling an even or odd number are equal.

EXPLORATION 4:

A. blue, 1 out of 4
   green, 1 out of 4
   red, 2 out of 4

B. Yes. There are two reds and only one blue on the spinner.

C. Answers will vary.

Explore Some More:

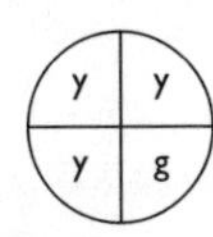


EXPLORATION 5:

A. 1 out of 8 for each number

B. Answers will vary. Each number has an equal chance because each is on the spinner one time.

C. Answers will vary. Chances of spinning an even number and odd number are equal.

## EXPLORATIONS 6-10

EXPLORATION 6:

A. Bag 1—Green, Green, Red, Red, Blue, Blue
   Bag 2—Blue, Blue, Blue, Red, Green, Purple
   Bag 3—Blue, Blue Blue, Green, Green, Yellow

B. Answers will vary.

Think and Write: The results may not exactly match the clue.

EXPLORATION 7:

A. purple—There are three purple cubes in the bag.
   white—There is only one white cube in the bag.
   blue and green—There are two of each in the bag.

B. white, green, green, blue, blue, purple, purple, purple

C. Answers will depend on cubes placed in the bag.

EXPLORATION 8:

A. Colors on spinner: Blue, Green, Red, Yellow
   Blue, Blue, Green, Yellow
   Green, Blue, Blue, Blue

B. Answers will vary. Results should begin to reflect chance.

Think and Write: The results may not exactly match the clue.

EXPLORATION 9:

A. Equally likely to roll an even and an odd number. Chances for even or odd are 3 out of 6.

B. Answers will vary.

Think and Write: The results may not match the prediction, but they should be close.

EXPLORATION 10:

A. More likely to roll an even sum because even + even = even; even + odd = even; odd + odd = odd.

B. Answers will vary.

Think and Write: The results may not match the prediction, but they should be close.

Explore Some More: More likely for product to be even because even x even = even; even x odd = even; odd x odd = odd.

## EXPLORATIONS 11-15

EXPLORATION 11:

A. Hannah will win because blue has a better chance of being spun.

Think and Write: No. Blue has a 2 in 4 chance of winning.

EXPLORATION 12:

A and B. Answers will vary.

EXPLORATION 13:

A. Joel has a better chance of winning because there are more odd sums.

B. Answers will vary.

Think and Write: There are more odd sums, so odd player wins.

Explore Some More: The even player will win because there are more even products.

EXPLORATION 14:

A. Ira and Amy have the same chance of winning.

B. The results should show an equal chance of winning. Color choice should not matter.

Think and Write: Chances are equal.

EXPLORATION 15:

A. Ben and Alex have an even chance of winning. The game is fair.

B. Answers will vary.

Explore Some More: The game is fair is players have an equal chance to win.

## EXPLORATIONS 16-20

EXPLORATION 16:

A. Answers will vary. There should be more blue cubes in the bag; yellow and green should be the same. Other students will have other combinations.

B. Answers will vary.

Explore Some More: The chance does not change, but the results may get closer to the prediction.

EXPLORATION 17:

A. 1. 28 students
2. 8 students
3. 108 students
4. 2 students; the symbol would stand for 1 student. It is not incorrect.

B. Gymnastics ○ ○ ○ ◖
Art ○ ○ ○
Running ○ ○
Cooking ○ ○ ◖
Basketball ○ ○ ◖

EXPLORATION 18:

A. Answer will be represented with the corresponding manipulative.

B. Power Man □ □ □
PC Ten □ □ □ □ □ □ □ □ □
Mountain Quest □ □ □ □ □ □ □ □
River Run □ □ □ □ □ □ □ □ □ □ □ □

EXPLORATION 19:

A. Description 2 matches the line plot.
Labels: chocolate, strawberry, vanilla

B. Answers will vary.

EXPLORATION 20:

A. red and green both came up four times; yellow and blue came up four times; yellow and blue results added together equal the results for red and green

B. Answers will vary; chance for each number is equal.

Think and Write: A line plot makes it easy to record data because you can keep track of the information as you record it. When you are finished you can easily read the data collected without having to make a graph first.

## EXPLORATIONS 21-25

EXPLORATION 21:

A. Tigers are having a losing season with average score fewer than 3 runs.

B. 1. Rangers—the nose
2. the eyes
3. More than 5—the mouth

EXPLORATION 22:

A. Answers will vary.

B. Everyone should have the same symbols unless students start school at different times.

EXPLORATION 23:

B. 1. 55
2. red
3. Any color is likely to come up. Each color has the same chance.

EXPLORATION 24:

B. A Dice Bar Graph

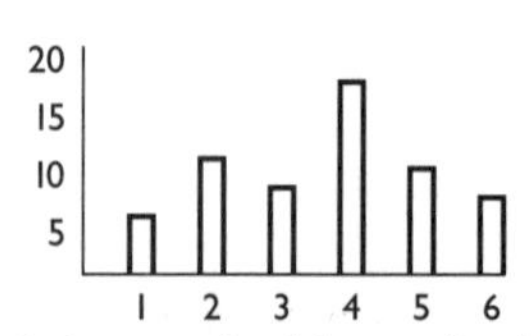


B. What number was rolled most often? Least often?
Which two numbers were rolled the same number of times?

EXPLORATION 25:

A. This will be answered using the cubes.

B. 1. Who read the most books?
2. Who read half as many books as Jerome?
3. Who read two fewer books than Keisha?

C. Bar graph reflects information in table.

Think and Write: The cube graph shows the exact numbers. the bar graph is measured along the axis.

## EXPLORATIONS 26-30

EXPLORATION 26:

A. 1. children's bikes
2. bike seats
3. Fred sold the same number of men's and women's bikes.

B.

EXPLORATION 27:

A.

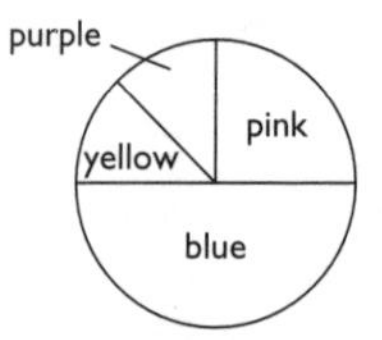


B. Blue is the largest section.
Yellow and purple together are about the same size as pink.
Blue—1/2, Pink—1/4, Yellow—1/8, Purple—1/8

Think and Write: The circle graph is a good way to display information because you can see what part a category is compared to the whole.

EXPLORATION 28:

A. Bar Graph

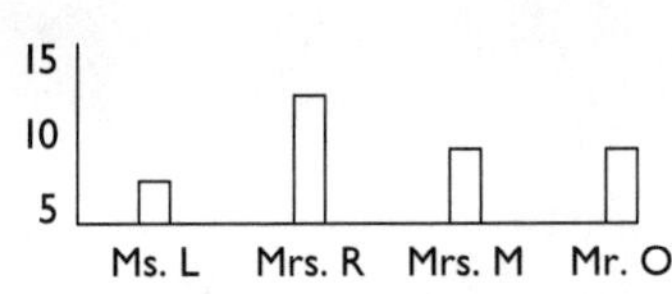


B.

Think and Write: Both work. The bar graph shows comparisons. The circle graph shows the parts fit in with the whole.

EXPLORATION 29:

A. Predictions will vary.
   6 ways to line up
   The cubes helped solve the problem because I could move them and record results.

B. 24 ways to line up

Think and Write: 6 numbers

EXPLORATION 30:

A. 9 outfits

B. 16 outfits

Think and Write: Making a list helps you solve this problem because it helps you keep track of the outfits you have created.

## EXPLORATIONS 31-35

EXPLORATION 31:

A. Samantha and Alexis's cubes equal Jamal and Henry's cubes
   Jamal and Henry both have 12.

B. mode is 7

Think and Write: Ideas about who will win will vary. Jamal and Elisa are ahead.

EXPLORATION 32:

A. The students should take the uneven stacks of cubes and move them to make the stacks even.

B. 3 is the average number of cubes.

Think and Write: Move the cubes to make even stacks to find the average.

EXPLORATION 33:

Answers will vary.

Think and Write: The yarn helps you find the average length because you divide the yarn into 4 equal pieces. The length of a piece is the average length.

EXPLORATION 34:

A. The average estimate is 12 cubes.
   The average is 10 cubes.

B. Answers will vary.

Explore Some More: Answers will vary.

EXPLORATION 35:

A.

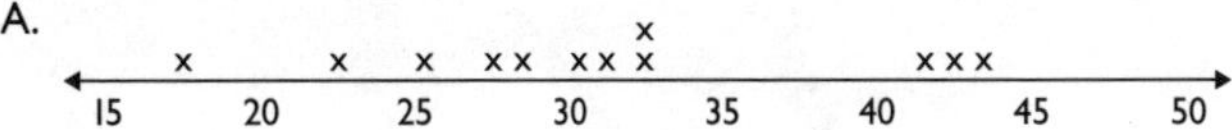


largest is 44, smallest is 18, 26 is the range
The line plot organized the information so that I could find the range.

B. range of estimates is 17, range of exact amounts is 14; the estimates have a larger range

Think and Write: The range tells you how far about the data is spread. If the range is 5 the data is close together. If the range of estimates is 35, the data is spread apart.

## EXPLORATIONS 36-40

EXPLORATION 36:

A.

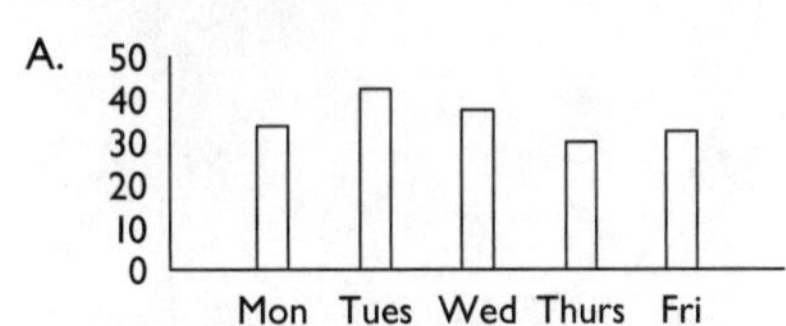


Temperatures for the Week

B. I put 38 close to the 40 mark on the line.
   The range is 10 degrees.
   32 degrees occurred more than once.
   The difference between the temperatures on Tuesday and Wednesday is 4 degrees.

EXPLORATION 37:

A. Description 3 best describes the line graph.

B. The temperatures started at 34 degrees, fell to 10 degrees, and began to increase, finally ending the week at 32 degrees.

EXPLORATION 38:

A. The graph tells how the temperature changes from day to day.

EXPLORATION 39:

A. 1. not shown
   2. Graph A
   3. Graph B

B. Description 1 could be shown in a bar graph.

EXPLORATION 40:

A. Answers will vary. Circle graph will show popularity of pizza and chicken.

B. Answers will vary.

## Circle Graph

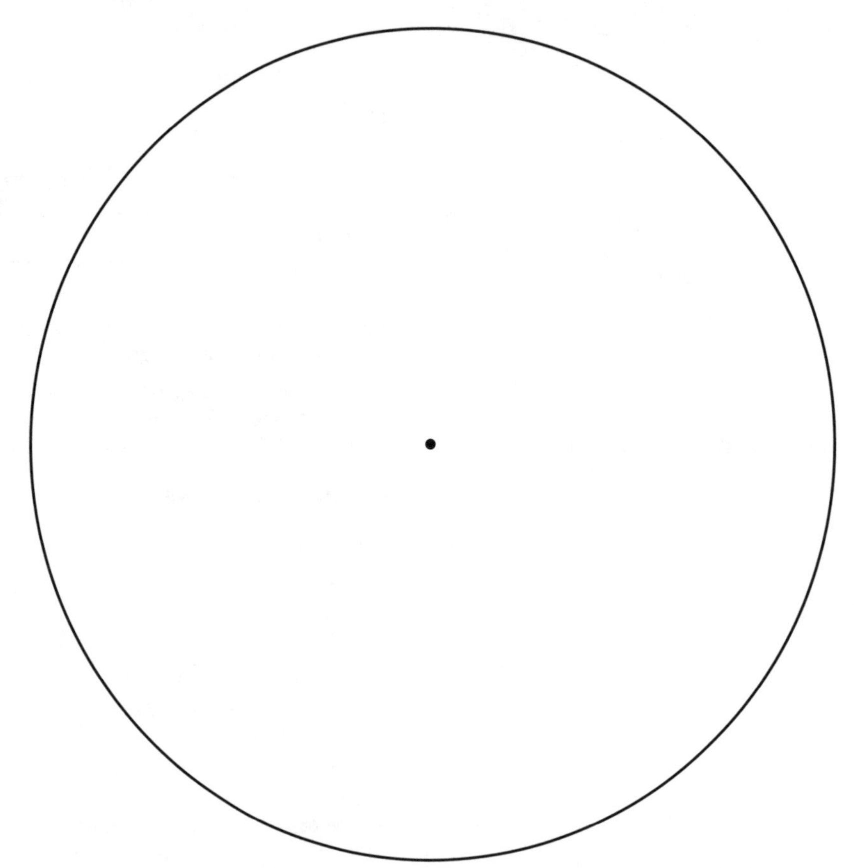

## Pattern for Regular Die

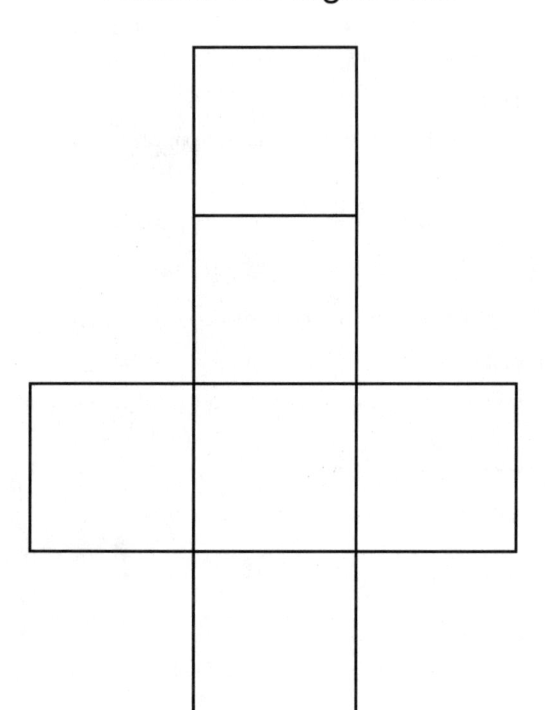

## Spinner

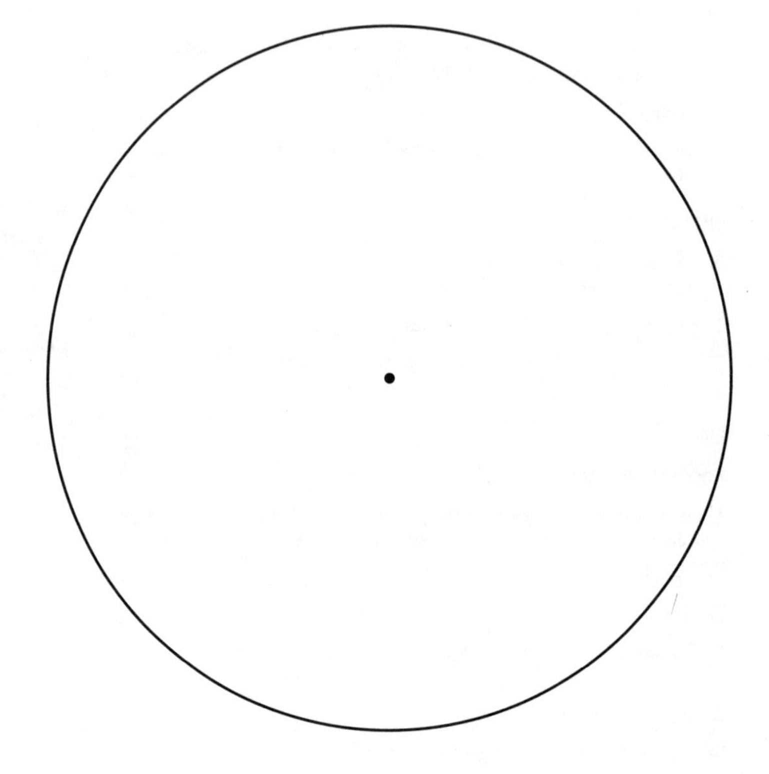

# NOTES ABOUT EXPLORATIONS 1-5

## MATH SKILLS AND UNDERSTANDINGS

- Explore concepts of chance through experiments with drawing cubes from a bag, rolling a die, and spinning a spinner
- Use tally marks in a table to record results of experiments
- Develop an understanding of possible outcomes in an experiment and determine what chance each has of occurring

## GETTING READY

| Heads | | | | |
|---|---|---|---|---|
| Tails | | | | |

Obtain a copy of *Cloudy with A Chance of Meatballs* by Judi Barrett, MacMillan Company, 1978. Also have available eight pennies. Draw a table showing the outcomes **heads** and **tails** on a chalkboard or an overhead transparency.

For the explorations, each pair of students will need one brown paper bag; 2 or more LacerLinks of each color: green, red, blue, yellow, and purple; one die numbered 1–6; and one transparent spinner. If transparent spinners are not available, students can use a pencil and paper clip as a spinner.

## INTRODUCING THE EXPLORATIONS

Read aloud *Cloudy With a Chance of Meatballs*. Discuss what *chance* means. Elicit ideas about the chance of something happening.

Have the students watch as you flip a penny. Ask them to identify how it lands—heads or tails. Show students how to use the tally marks to record the results. Flip the penny a few more times and record the results. Then ask, **If the penny lands on heads, will it land on tails the next flip?** Elicit from the students that there are two possible results or outcomes: heads and tails. Develop the concept that each of these outcomes has an equal chance of happening. Say, **There is one chance out of two that the penny will land on heads. What can you say about the chance that the penny will land on tails?** (1 out of 2) Develop the concept that the previous outcomes do not affect the next outcome.

Ask, **If you flip a penny 20 times, do you think the results will be 10 heads and 10 tails?** Divide the class into groups of eight, and give each group a penny and a sheet of scratch paper. Have students take turns flipping the penny for a total of 20 flips. Tell them that each flip is called a *trial*. Have them use a tally mark to record each trial. Then help the groups compare their results.

## TALKING AND WRITING ABOUT THE EXPLORATIONS

To help the students reflect upon what they have discovered during the explorations, you can ask questions for discussion or journal writing: **How did you find out what could happen, or the possible outcomes? How did you find the chance of something happening? Was the chance of something happening the same as what really happened in your experiments? Why or why not?**

## EXTENSION IDEAS

Students can look in the newspaper to find articles that use words such as *chance, prediction,* and *outcomes*. Have the students share the articles in class.

Name ..........................................................................

Use LacerLinks, a paper bag, and a pencil.
Record your work.

A. Jacob put 1 green cube, 1 red cube, 1 yellow cube, and 1 blue cube in a paper bag.

How many cubes are green? _____

How many cubes are red? _____

How many cubes are yellow? _____

How many cubes are blue? _____

B. Jacob shakes the bag. Greta draws one cube out of the bag without peeking. Her chance of getting green is 1 chance out of 4. What is her chance of getting each of the other colors?

green __1__ chance out of __4__

red _____ chance out of __4__

yellow _____ chance out of _____

blue _____ chance out of _____

Is Greta more likely to draw one color than another? Why or why not?

C. Try Jacob's experiment. Put the same cubes in the bag as he did. Draw one cube out of the bag. Use a tally mark to record the color. Put the cube back in the bag and shake it. Repeat 20 times in all.

| | |
|---|---|
| green | |
| red | |
| blue | |
| yellow | |

Think and Write

Compare your results with a partner's results. Are they alike? Are they different?

Name ................................................................................

Use LacerLinks, a paper bag, crayons or markers, and a pencil. Record your work.

A. Greta put 1 green cube, 1 red cube, 1 purple cube, and 1 blue cube in a paper bag. If Jon draws a cube out of the bag, what is his chance of getting each color?

green _____ chance out of _____ red _____ chance out of _____

purple _____ chance out of _____ blue _____ chance out of _____

B. If Greta wants to change the chance of drawing a green cube to be 2 out of 4, what could she do? Color the cubes to show the change.

☐ ☐ ☐ ☐

C. Jon put 6 new cubes in the bag. Read the clues to figure out which cubes he chose. Color the cubes to show what he chose.

- The chance of drawing a red cube is 2 out of 6.
- The chance of drawing a blue cube is 1 out of 6.
- The chance of drawing a purple cube is 3 out of 6.

☐ ☐ ☐ ☐ ☐ ☐

Explore Some More

Take six cubes and hide them from a friend. Give your friend clues about the chance of drawing each color. See if your friend can tell you which cubes you chose. Now put the cubes in the bag. Draw out one cube, record the color, and put it back in the bag. Do this 20 times. What happens? Are you surprised?

Name ............................................................

Use one die, and a pencil.
Record your work.

A. Frank and Marie are rolling a die. Frank says that it is easier to roll an even number than an odd number. Do you agree? Why or why not?

B. Each number is one possible outcome. What is the chance of getting each number when you roll a die?

| Number | Possible Outcome | Chance |
|---|---|---|
| 1 | | out of |
| 2 | | out of |
| 3 | | out of |
| 4 | | out of |
| 5 | | out of |
| 6 | | out of |

What is the chance of getting an even number? _____ out of _____

What is the chance of getting an odd number? _____ out of _____

C. With a partner, do Frank and Marie's experiment. Roll a die 36 times. Take turns. Each time you roll the die, record whether you get an even number or an odd number. Use tally marks.

| even number | |
|---|---|
| odd number | |

Look at the results of your experiment. Is there a better chance of rolling an even number than an odd number? Why or why not?

 

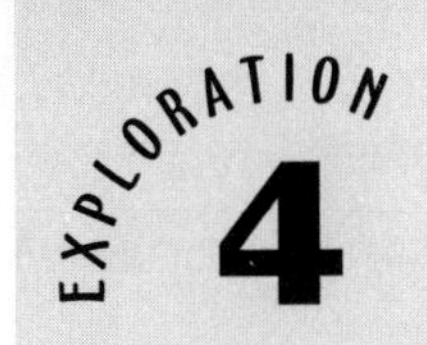


Name ____________________

Use a transparent spinner, crayons or markers, and a pencil. Record your work.

A. Tress and Tempe are spinning a spinner like this one. Each time they spin, the spinner can land on one color. Each color is one **possible outcome.** What is the chance of landing on each color?

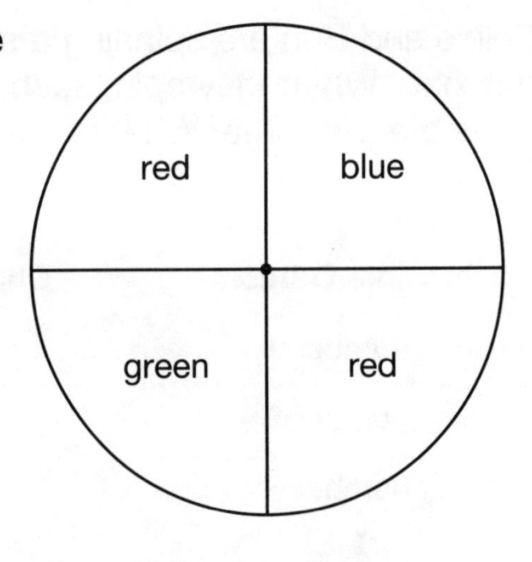


| Possible Outcome | Chance |
|---|---|
| blue | |
| green | |
| red | |

B. Tress says that there are twice as many chances that the spinner will land on red as on blue. Do you agree? Why or why not?

C. With a partner, spin the spinner 30 times. Take turns. Each time you spin, record what color the spinner lands on. Use tally marks.

| | |
|---|---|
| blue | |
| green | |
| red | |

Look at the results of your experiment. Did the spinner land on red twice as many times as on blue or on green?

Explore Some More

Design a spinner so that there is a better chance of landing on yellow than on green.

## EXPLORATION 5

Name ____________________

Use a transparent spinner and a pencil.
Record your work.

A. Felice and Ben are spinning a spinner like this one. When they spin, each number is a **possible outcome.** What is the chance of landing on each number?

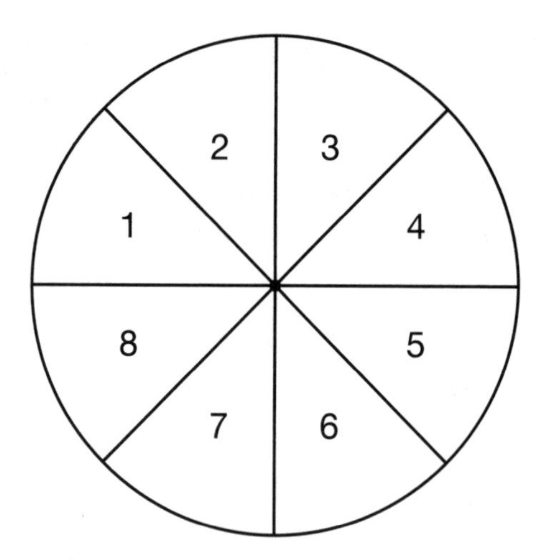


| Possible Outcome | Chance |
| --- | --- |
| number 1 | |
| number 2 | |
| number 3 | |
| number 4 | |
| number 5 | |
| number 6 | |
| number 7 | |
| number 8 | |

B. With a partner, spin the spinner 30 times. Take turns. Each time you spin, record what number the spinner lands on. Use tally marks.

| | |
| --- | --- |
| 1 | |
| 2 | |
| 3 | |
| 4 | |

| | |
| --- | --- |
| 5 | |
| 6 | |
| 7 | |
| 8 | |

What did your experiment show? Does each number have an equal chance of the spinner landing on it? How do you know?

C. Look at the results of your experiment.
How many times did you spin

an even number? _____ an odd number? _____

 

# NOTES ABOUT EXPLORATIONS 6-10

## MATH SKILLS AND UNDERSTANDINGS

- Explore possible outcomes of rolling dice, drawing cubes from a bag, and spinning a spinner as being equally likely, more likely, less likely, certain, or uncertain
- Use tally marks to record results of experiments
- Design a game to determine the conditions of equally likely, more likely, less likely, certain, or uncertain
- Understand that there are ways to find out why some outcomes are more likely than others

## GETTING READY

Write the following on index cards:

You will also need a six-foot length of yarn and clothespins.

For the explorations, each pair of students will need one regular die numbered 1-6, transparent spinners, and LacerLinks.

## INTRODUCING THE EXPLORATIONS

Have two students come to the front of the room and hold the yarn between them so that it is stretched end to end. Clothespin the card labeled *Uncertain* at the left end, and the card labeled *Certain* at the right end of the yarn. Explain that the likelihood of something happening increases as you move along the line from uncertain to certain.

Display the statements on the cards and ask volunteers to place them on the line where they think it makes the most sense. Ask, **Why did you place the card there? Can you explain your thinking?**

Divide the class into groups of four. Assign each group a label: *certain, uncertain, less likely, more likely.* Have each group make up several statements to go with its label:

It is certain that ____________________

It is uncertain that ____________________

It is more likely that ________ will happen than ________ .

It is less likely that ________ will happen than ________ .

Have students share completed statements with the rest of the class. Discuss whether the statements make sense.

## TALKING AND WRITING ABOUT THE EXPLORATIONS

To help the students reflect upon what they have discovered during the explorations, you can ask questions for discussion or journal writing: **How did you determine whether something was more likely, less likely, or equally likely to happen? Was the likelihood of something happening the same as what actually happened in your experiments? Why or why not?**

## EXTENSION IDEAS

Students can watch a weather report and determine how many times the phrases less likely or more likely are used. Then have the students compare what was forecasted to what actually happened.

Name ............................................................

Use LacerLinks, a paper bag, crayons or markers, and a pencil. Record your work.

A. Three bags are shown below. Each bag has 6 cubes in it. Read the clue under each bag. Then color the cubes in the bag to match the clue.

**Bag 1**

All colors have an equal chance to be drawn.

**Bag 2**

Blue is most likely to be drawn.

**Bag 3**

Yellow is least likely to be drawn.

B. Experiment to see if the actual results match the clue. Get 6 cubes that match the colors you chose for one of the bags above. Write the colors in column 1 on the chart. Put the cubes in a paper bag. Draw one cube out of the bag. Use a tally mark to record the color. Put the cube back in the bag. Shake the bag. Draw 24 times in all.

| Color | Results |
|---|---|
| | |
| | |
| | |
| | |
| | |
| | |

Think and Write

Did your results match the clue? Why or why not?

Name ............................................................

Use LacerLinks, a paper bag, and a pencil.
Record your work.

A. Shelby puts 3 purple cubes, 2 blue cubes, 2 green cubes, and 1 white cube in a paper bag. She shakes it up. Trinh draws one cube out of the bag.

What color is Trinh most likely to draw?

Why do you think so?

What color is she least likely to draw?

Why do you think so?

What colors have an equal chance of being drawn?

Why do you think so?

B. Take Shelby's cubes. Put them in order on the line to show their chance of being drawn from her bag. Color the boxes to show where you put the cubes.

least likely ⟷ most likely

□ □ □ □ □ □ □ □

C. Choose 8 cubes to put in a bag. Color the boxes to show which cubes you chose.

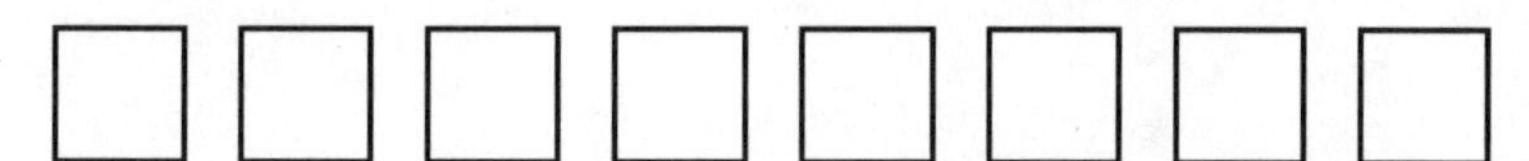

Put the cubes in order on the line to show their chance of being drawn from a bag. Color the boxes to show where you put the cubes.

least likely ⟷ most likely

□ □ □ □ □ □ □ □

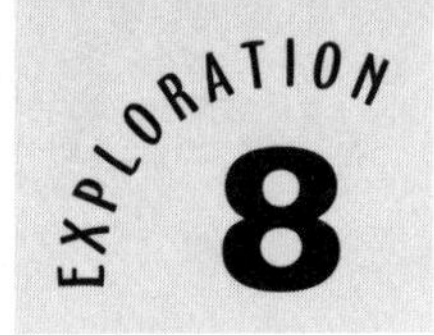


Name ............................................................

Use a transparent spinner, crayons or markers, and a pencil. Record your work.

A. Two blank spinners are shown below. Think about the clue given under each spinner. Color the spinner to match the clue.

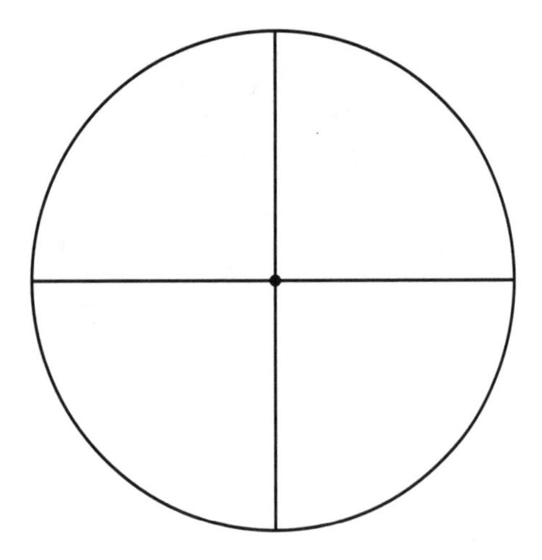

All colors are equally likely to be spun.

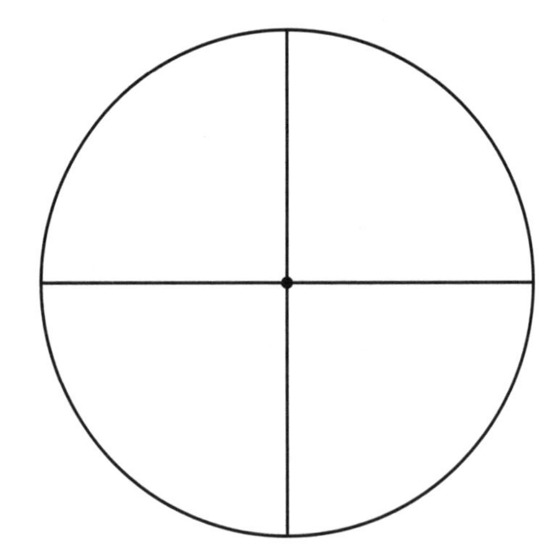

Blue is more likely to be spun.

B. Choose one of the spinners. Do an experiment to see if the actual results will match the clue. Place a transparent spinner on the spinner. Spin the spinner. Record the color. Spin 24 times altogether.

| Color | Results |
| --- | --- |
| | |
| | |
| | |
| | |

Think and Write

Did your results match the clue? Why or why not?

Name ____________________

Use a die and a pencil.
Record your work.

A. Marlene rolled a 3 on a die. Then she rolled a 4. After that she rolled a 2. Marlene wondered what the chance would be of rolling an even number on the die.

What are the chances that Marlene will get an even number on her next roll? Will she be more likely to roll an even number? Will she be less likely? Or is she equally likely to roll an even number as an odd number? Why do you think so?

B. Test your prediction by rolling a die 20 times to see what the results will be. Roll the die. Record whether the number is even or odd.

| Even | Odd |
| --- | --- |
| | |

Think and Write

Do your results match your prediction? Why or why not?

Name ____________________

Use dice and a pencil.
Record your work.

A. Marlene rolled one die. She discovered that there was an equal chance of rolling an even number and odd number. Jerome wondered what the chance would be of getting an even sum by rolling two dice and adding the two numbers. Would he be more likely or less likely to roll an even sum than an odd sum? Or would an even sum and an odd sum be equally likely? Why do you think so?

B. Test your prediction. Roll two dice 30 times. On the table below, record whether the sum of the numbers is even or odd.

| Even | Odd |
|---|---|
| | |

Think and Write

Do your results match your prediction? Why or why not?

Explore Some More

What if you were to find the product of two numbers on the dice instead of the sum? Would the results be the same?

  

# NOTES ABOUT EXPLORATIONS 11-15

## MATH SKILLS AND UNDERSTANDINGS

- Explore concepts of chance through experiments with color and number spinners and LacerLinks
- Investigate chance by playing spinner games, drawing cubes from a bag, and constructing a die

## GETTING READY

Use transparent spinners to make two spinner circles like the ones shown.

**Spinner One**

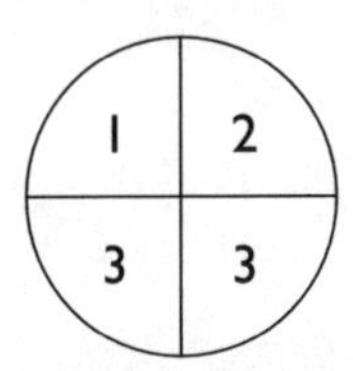


**Spinner Two**

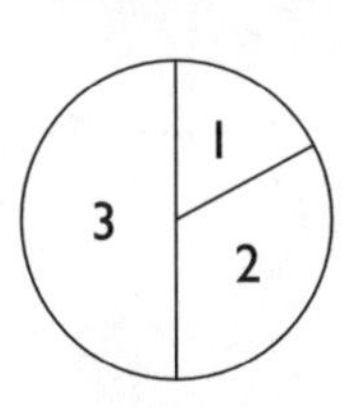


For the explorations, each pair of students will need a transparent spinner, crayons or markers, red and blue LacerLinks, paper pattern for a die (page x), and a brown paper bag.

## INTRODUCING THE EXPLORATION

Show the students Spinner One. Tell them that three students are playing a game. Each player has a number: 1, 2, or 3. The players get a point when the spinner lands on their number. Divide the class into groups of three. Have each group member choose a number. Elicit the idea that number 3 has a better chance of winning than 1 or 2. Develop the idea that there is twice as much chance of spinning a 3 as a 2 or 1. Ask, **Is this spinner game fair to player 1 and 2? Why or why not?**

Show the students Spinner Two. Ask, **Is this is a fair spinner for all three players? Why or why not? How could you make the spinner fair?** If the spinner is fair, each player has an equal chance of winning. Have the groups share their ideas about how to make the spinner fair.

## TALKING AND WRITING ABOUT THE EXPLORATIONS

To help the students reflect upon what they have discovered during the explorations, you can ask questions for discussion or journal writing: **How did you determine whether something was fair or unfair? How did the experiments help you find out if the game was fair or unfair? What did you make sure to do when making a fair game? How do you know when a game is fair? If someone wins 5 games in a row, is the game is unfair? Why or why not?**

## EXTENSION IDEAS

Give the students two blank spinner circles (page x) and ask them to make a fair spinner and an unfair spinner. Ask the students to write an explanation of why one spinner is fair and one spinner is not.

Ask the students if they have ever been to a carnival, a state fair, or an amusement park and played a game of chance. Ask the students if the game was a fair game. How can students tell whether a game is fair?

Name

Use a transparent spinner, crayons or markers, and a pencil.
Record your work.

A. Hannah made up a spinner game to play with Jim. Every time they spin blue, Hannah will get a point. If they spin red, Jim will get a point. If they spin yellow, both of them will get a point. The first person to get 10 points wins.

This is the spinner Hannah made for the game.

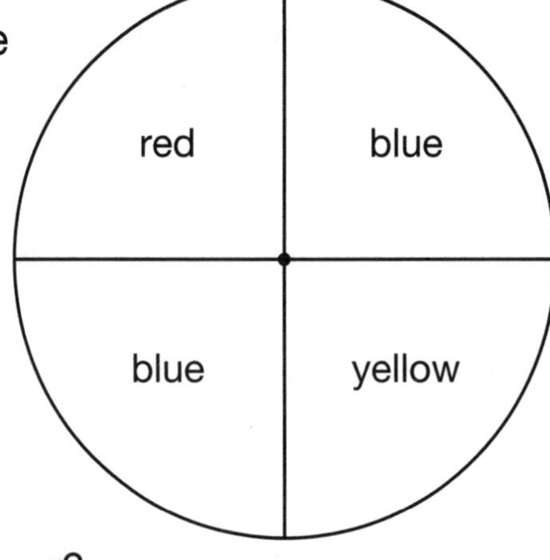


Who do you think will win the game?

Why do you think so?

B. Test your prediction by playing Hannah's game with a partner. Color the spinner. Take turns spinning. The first player to spin blue gets blue. The other player gets red. Use tally marks to record your points. Keep playing until someone has 10 points.

| Color | Player 1 | Player 2 |
|---|---|---|
| blue | | |
| red | | |
| yellow | | |

Think and Write

Is the game fair? Why or why not?

Name ..........................................................................................................

Use a transparent spinner, crayons or markers, and a pencil. Record your work.

A. Color Hannah's game spinner in a way that will give Hannah and Jim an equal chance of winning. Write the rules for them to use.

Rules:

The first player to get _____ is the winner.

B. Try your game with a partner. Use tally marks to record your results.

| | Player 1 | Player 2 |
|---|---|---|
| | | |
| | | |
| | | |

Think and Write

Is your spinner game fair? How do you know? How did you make the spinner fair for both players?

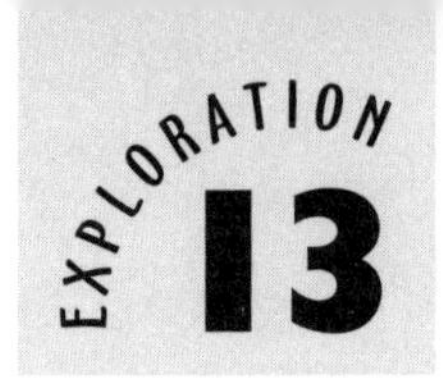


Name ............................................................

Use a transparent spinner and a pencil.
Record your work.

A. Heather and Joel are playing a game with the spinner. Each player takes a turn and remembers the number. They add the two numbers together. If the sum is even, Heather gets a point. If the sum is odd, Joel gets a point. Does either player have a better chance of winning?

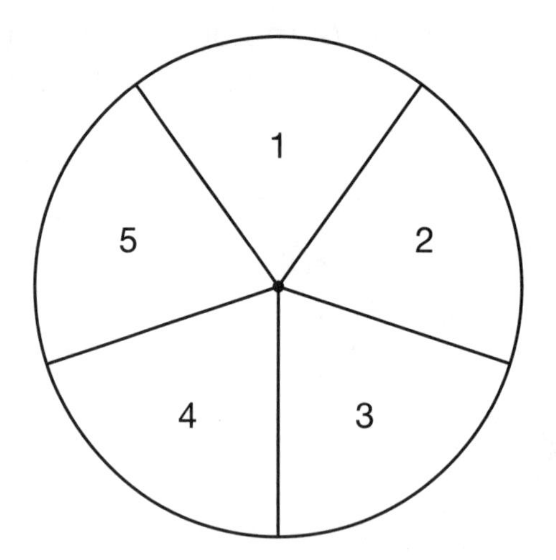


Why or why not?

B. With a partner, play the game. Who will get a point if the sum is even? Who will get a point if the sum is odd? Record who wins each trial.

| Player name | Even | Odd |
|---|---|---|
| | | |
| | | |

Think and Write

Who won? Are there more even sums or odd sums? Which player would you rather be? Why?

Explore Some More

Use the same spinner and play the game with a partner. This time, multiply the two numbers. If the product is even, Player A gets a point. If the product is odd, Player B gets a point. Do you think someone has a better chance of winning? Why or why not? On another paper, record your results for 25 trials.

EXPLORATION **14**

Name ________________

Use 2 brown bags, 2 red LacerLinks, 2 blue LacerLinks, and a pencil.

A. Ira and Amy are playing a game. They each have a bag with a red cube and a blue cube in it. At the same time, each of them pulls out a cube. If the cubes are the same color, Ira gets a point. If the cubes are different colors, Amy gets a point. Ira thinks he will win. What do you think? Does either Ira or Amy have a better chance of winning, or are their chances the same?

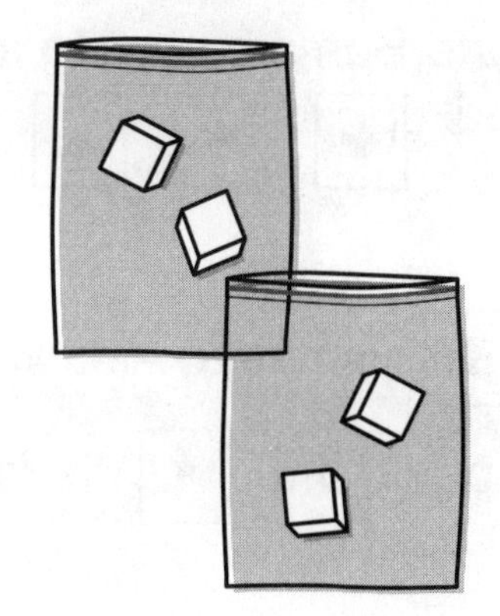

B. With a partner, play the game. Each player puts a red and a blue cube in a bag. Decide who gets a point if the cubes are the same color and who gets a point if the cubes are different. At the same time, each of you pull a cube out of your bag. Record 25 trials.

| Name of player | Same color | Different color |
|---|---|---|
| | | |
| | | |

What are the results after 25 trials? Did either player have a better chance of winning?

If you could choose to be the player who got a point for the same color or the different color, which player would you be? Why?

Think and Write

What are the possible outcomes for drawing two colors the same and two different colors. What are the chances of drawing two different colors?

Name ........................................................................................................

Use a paper pattern for a die, scissors, tape, and a pencil.
Record your work.

A. Alex and Ben are playing a game with special dice. The dice are marked ✚ ✚ ▲ ▲ ★ ★

If a ✚ is rolled, Alex will get a point. If a ▲ is rolled, Ben will get a point. If a ★ is rolled, neither Alex nor Ben gets a point.

Who will win the game?

Is the game fair? Why or why not?

B. Make up a dice game that is fair. You can design your own die and make up the rules for your game. All players must be equally likely to win the game.

Dice Game Rules

______________________________

______________________________

______________________________

______________________________

______________________________

______________________________

Show how your die is marked.

Top: ____________

Bottom: ____________

Side 1: ____________

Side 2: ____________

Side 3: ____________

Side 4: ____________

Explore Some More

Try out your game with a friend. Is your game fair? How do you know?

 

## MATH SKILLS AND UNDERSTANDINGS

- Explore concepts of collecting data and recording data using tally marks
- Construct a line plot by spinning a spinner and interpret results
- Construct a pictograph using LacerLinks to display given data, record and interpret results
- Use deductive reasoning to interpret a line plot

## GETTING READY

Cut out enough graphs from the newspaper so that each group of four can have at least two graphs. Make a transparency of each kind of graph. *USA Today* is a good source.

For the explorations, each pair of students will need a brown paper bag; red, blue, green, and white LacerLinks; a die; crayons or markers.

## INTRODUCING THE EXPLORATIONS

Ask the students why information is shown in graphs. Elicit that graphs help show information visually. Divide the class into groups of four. Distribute the graphs from the newspaper. Ask, **What do the graphs show?** List responses on the chalkboard. Display a transparency of each graph distributed to the students. Ask, **What does this graph look like? What is the name of this kind of graph?** Elicit that a bar graph has bars, a pictograph shows the data pictorially, and a circle graph encloses the data in a circle. Ask, **What do all of these graphs have in common? How are they alike? How are they different?** Have the students brainstorm the characteristics of a graph. Record their ideas on a web.

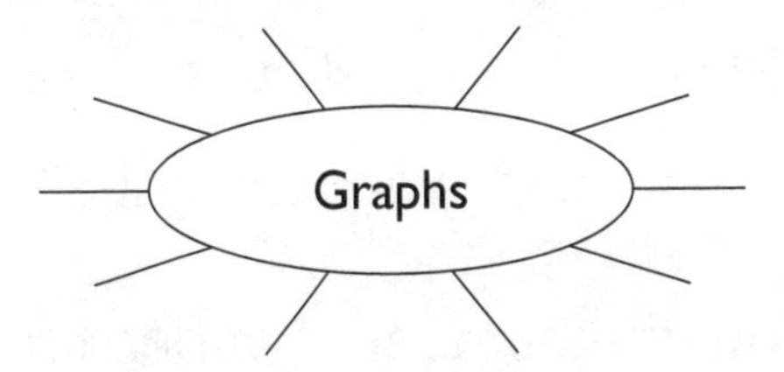


Ask the students to name the different kinds of graphs.

## TALKING AND WRITING ABOUT THE EXPLORATIONS

To help the students reflect upon what they have discovered during the explorations, you can ask questions for discussion or journal writing: **Why are tally marks a good way for recording collected data? Why are line plots a good way to record data? Why would you record data on a line plot instead of using tally marks? How is a pictograph read?**

## EXTENSION IDEAS

Ask students to identify the kinds of shoes they are wearing—tennis shoes, shoes with straps, slip-on shoes, boots. Have them decide on a way to categorize the shoes, such as shoes with laces and shoes without laces. Write the category labels on index cards and place them on the floor to form two rows. Have students wearing shoes with laces take off one shoe and place it in the row. Do the same with students wearing shoes without laces. Have students identify the bar graph and think of a good title for it. Divide the class into four groups. Label index cards: *tally marks, line plot, pictograph, circle graph.* Give a card to each group. Each group must record the shoe data according to the label on the index card. Have the students display and discuss their work.

Name ____________________

Use a paper bag, LacerLinks, and a pencil.
Record your work.

A. With a partner, put 10 cubes in a bag. Use these clues to decide which cubes to use.

- Blue is more likely to be drawn than red.
- Yellow and green are equally likely to be drawn.

Write the number of cubes for each color.

blue _____ red _____ green _____ yellow _____

Based on the cubes that you put in the bag, answer these questions.

What are the chances of drawing a blue cube? __________

What are the chances of drawing a green cube? __________

Do you think other pairs of students will have the same number of blue cubes that you and your partner have?

Why or why not?

B. Now try the experiment. Use the cubes that you put in your bag. Pull one cube out and record the color. Replace the cube and shake the bag. Repeat 20 times. Use tally marks to record your results.

| Color | Tally marks | Number |
|---|---|---|
| | | |
| | | |

Look at your results. How do your results match your predictions?

Did blue cubes get drawn more often than red?

## Explore Some More

Trade bags with another pair of students and repeat the experiment. Compare your results. Then put your data together. Did the chances of drawing a certain color change with more data?

 

EXPLORATION **17**

Name ____________________

Use LacerLinks, crayons or markers, and a pencil.
Record your work.

A. The After-School club made a graph. It shows how many students joined each group.

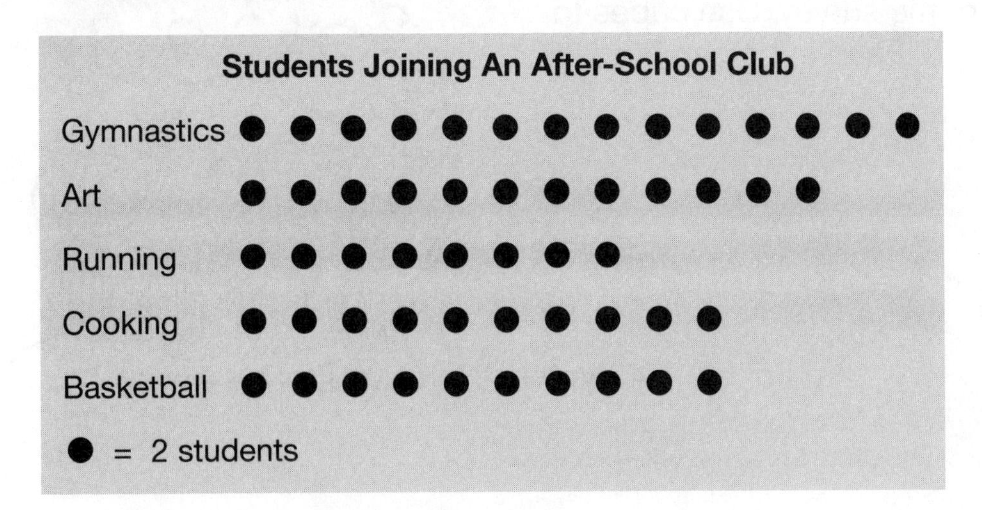


Look at the pictograph and answer the questions. You may want to use cubes to make the graph and then answer the questions.

1. How many students are in the gymnastics group?

2. How many more students are in the art group than the running group?

3. How many students joined groups in all?

4. What does ● mean in this graph?
   Why would ◖ be incorrect in this graph?

B. What would the graph look like if ● stood for 8 students? Make the graph in the space below.

Explore Some More

Write two questions that can be answered by looking at the graph you made. See if a partner can answer your questions.

Name ..................................................

Use LacerLinks and a pencil.
Record your work.

A. Anna and Juan are taking a survey of their classmates' favorite video games. Here are the results of the survey. Use cubes to represent the data. Each cube can stand for 3 votes.

| | | |
|---|---|---|
| Power Man | Blue | 12 votes |
| PC Ten | Red | 36 votes |
| Mountain Quest | Green | 24 votes |
| River Run | White | 48 votes |

B. What would your graph look like if each cube stood for 4 students? Make the pictograph in the space below.

| | |
|---|---|
| Power Man | |
| PC Ten | |
| Mountain Quest | |
| River Run | |

C. Write two questions that can be answered by looking at your pictograph. See if a partner can answer your questions.

1. ______________________________

2. ______________________________

Name ..............................................................................

Use a pencil.
Record your work.

A. This line plot shows data about flavors of ice cream. Read each statement below. Which one describes the data shown on the line plot? Use the information to label the line plot with the flavors.

Favorite Ice Cream

| | | |
|---|---|---|
| X | | |
| X | X | X |
| X | X | X |
| X | X | X |
| ______ | ______ | ______ |

1. More people like chocolate ice cream than strawberry ice cream. More people like strawberry ice cream than vanilla ice cream.
2. More people like chocolate ice cream. The same number of people like strawberry and vanilla ice cream.
3. More people like chocolate ice cream than vanilla ice cream. More people like vanilla ice cream than strawberry ice cream.

B. Look at the line plot below. Write a statement to describe the data in the line plot. Use the information to label the line plot.

| | | | | |
|---|---|---|---|---|
| | | | | X |
| | | | | X |
| | X | X | | X |
| X | X | X | X | X |
| ______ | ______ | ______ | ______ | ______ |

Think and Write

Is it easier to record data on a line plot or use tally marks?

## EXPLORATION 20

Name ..........................................................

Use a die and a pencil.
Record your design.

A. Erica and Ronald have been spinning the spinner shown below. They have recorded their results on a line plot. After 12 spins, the line plot looks like this:

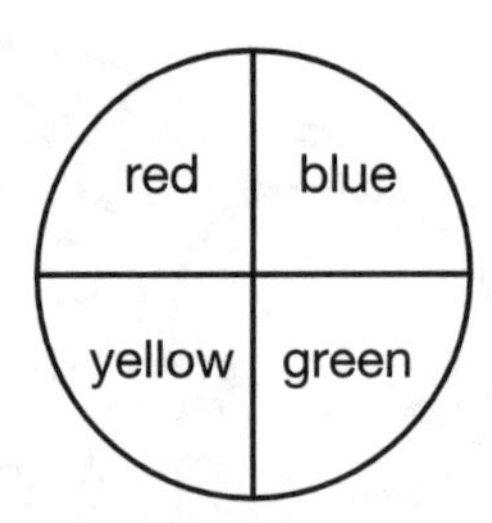


| X | | X | |
|---|---|---|---|
| X | | X | |
| X | X | X | X |
| X | X | X | X |
| Red | Yellow | Green | Blue |

What are the results for each color? Describe the data in your own words.

What two colors together have the same results as the other colors?

B. With a partner, roll the die 40 times Record the number that comes up with each roll. Put an X above the number on the line plot.

Before you begin, predict the results. Do you think a certain number will come up most often? Why or why not?

| 1 | 2 | 3 | 4 | 5 | 6 |
|---|---|---|---|---|---|

Think and Write

How does a line plot make it easy to record data and read data?

## MATH SKILLS AND UNDERSTANDINGS

- Explore concepts of displaying data by interpreting a facial glyph about a softball team
- Explore concepts of representing data by constructing a glyph about the school
- Explore concepts of interpreting data represented in a bar graph
- Explore concepts of constructing a bar graph using data represented by tally marks in a table

## GETTING READY

Have available worksheets showing a circle that measures 5 inches in diameter or a white paper plate and markers or crayons.

For the explorations, each pair of students will need crayons or markers, and LacerLinks of assorted colors. For Exploration 21, discuss finding an average and be sure students understand the concept and how it is calculated.

## INTRODUCING THE EXPLORATIONS

Ask the students to write the answers to the following questions on a piece of paper.

What is your eye color?

Is your favorite subject reading, mathematics, or science?

Do you have a pet?

Tell the students they are going to display their answers in a glyph. A glyph is a way of showing data. It uses a key. Facial glyphs use data to make a face. Give the students the 5-inch circle or a paper plate. Use the overhead projector to display the key and have students use it to fill in one feature at a time on their circles. Students can exchange glyphs and use them to describe each other's eye color, favorite subject, and pet.

| Eyes | Brown ○ ○ | Blue ⊙ ⊙ | Hazel/Green |
|---|---|---|---|
| Nose | Reading △ | Mathematics ● | Science ■ |
| Mouth | Yes | No | |

## TALKING AND WRITING

To help the students reflect upon what they have discovered during the explorations, you can ask questions for discussion or journal writing: **How do glyphs help show data? When you are constructing a bar graph, how do you decide the scale for the numbers? How do you decide if you are going to make the graph horizontal or vertical? How does a key help you interpret a glyph?**

## EXTENSION IDEAS

Ask the students to bring in weather maps from the newspaper. Tell them that a weather map is an example of a glyph. Have them make up a key to read a weather map.

Name ..............................................................................................

Use crayons or markers and a pencil.
Record your work.

A. A glyph is a way to show data. Randy and Renee saw these glyphs about two fourth-grade softball teams. Randy said, “Wow, the Tigers are having a tough time.” What made Randy say this? Use the key to help you answer.

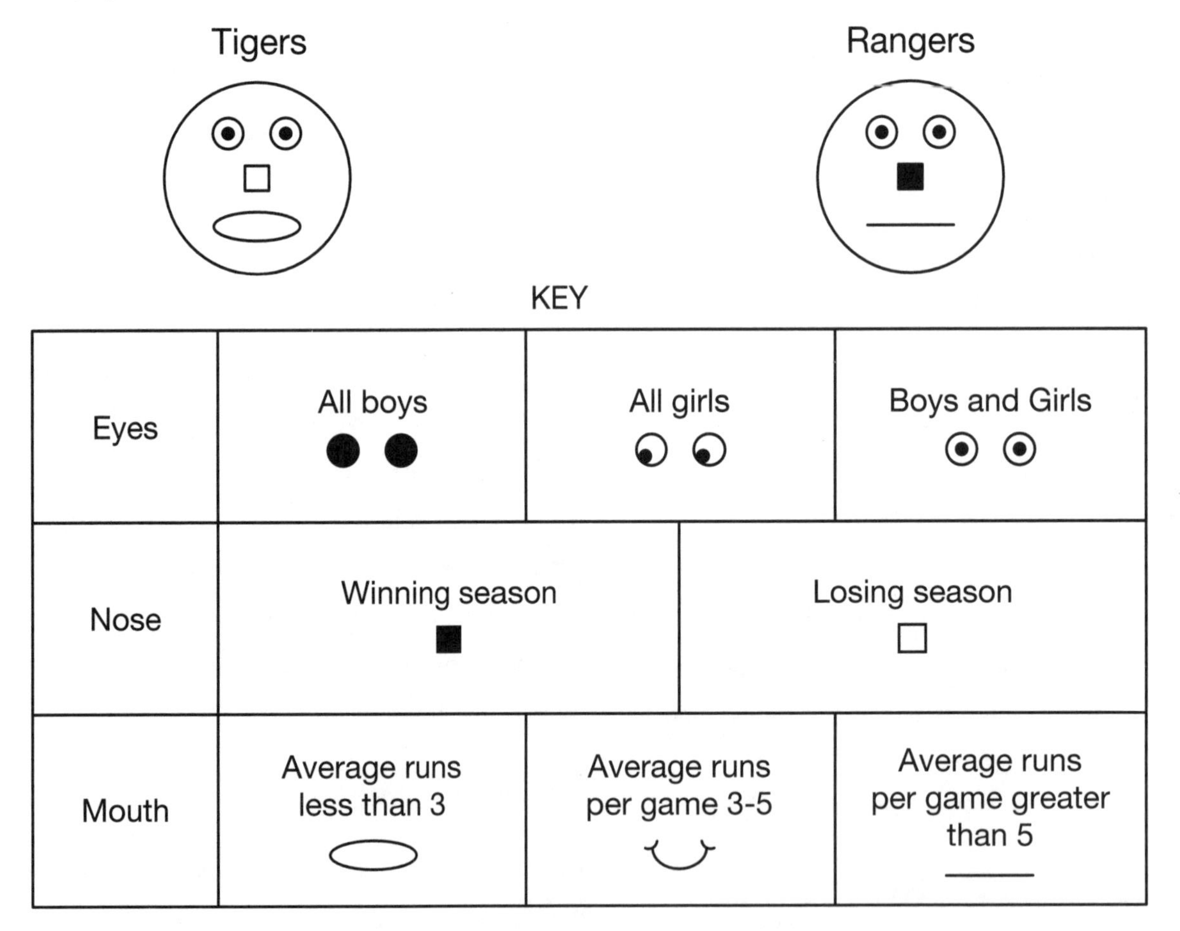


KEY

| | | | |
|---|---|---|---|
| Eyes | All boys | All girls | Boys and Girls |
| Nose | Winning season | | Losing season |
| Mouth | Average runs less than 3 | Average runs per game 3-5 | Average runs per game greater than 5 |

B. Look at the glyphs. Use the key to answer the questions.

1. Which team is having a winning season? How do you know?

2. What part of the face tells you who is on each team?

3. How many runs do the Rangers average per game? How do you know?

Think and Write

Write a question that you can answer by looking at the glyphs. See if a partner can answer your question.

Name ________________________________

Use crayons or markers, a sheet of paper, and a pencil.
Record your work.

A glyph is a graph that displays information.
Symbols are used to record data.
A key helps you read the symbols.

A. Use the symbols in the key below. On another paper, make a facial glyph that tells about your school.

KEY

| | | | |
|---|---|---|---|
| Eyes | Less than 250 students ◐ ◐ | Between 250 and 300 students ◓ ◓ | Greater than 300 students ⊙ ⊙ |
| Nose | Less than 12 teachers □ | 12 to 20 teachers △ | Greater than 20 teachers ▲ |
| Mouth | Starts before 8:30 a.m. 〰 | Starts between 8:30 and 9:00 a.m. ⬭ | Starts after 9:00 a.m. ◡ |

B. Get together with a small group. Compare your glyphs. How are they alike? Did you use the same symbols?

Explore Some More

Make a new key for a glyph. Share your key with a partner.
Ask your partner to use the key to draw a glyph.
See if you can interpret the glyph.

Name ........................................

Use a pencil.
Record your work.

A. Rusty and Ryan are experimenting with spinning a spinner like the one to the right. They used tally marks to record the spins. Then they put the data in a bar graph.

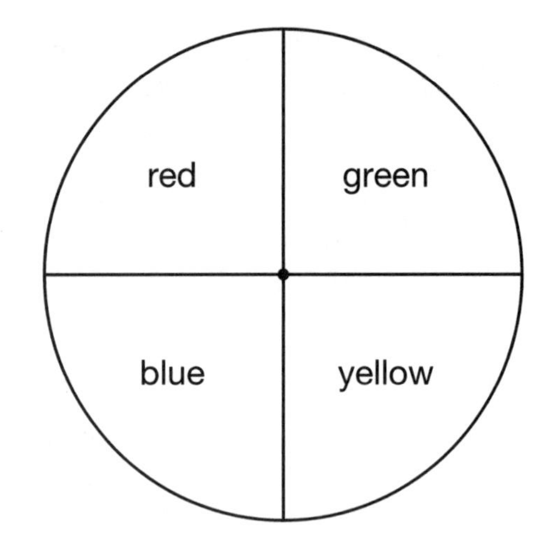


| | |
|---|---|
| Red 卌 卌 ||| | Yellow 卌 卌 卌 卌 |
| Blue 卌 | Green 卌 卌 卌 || |

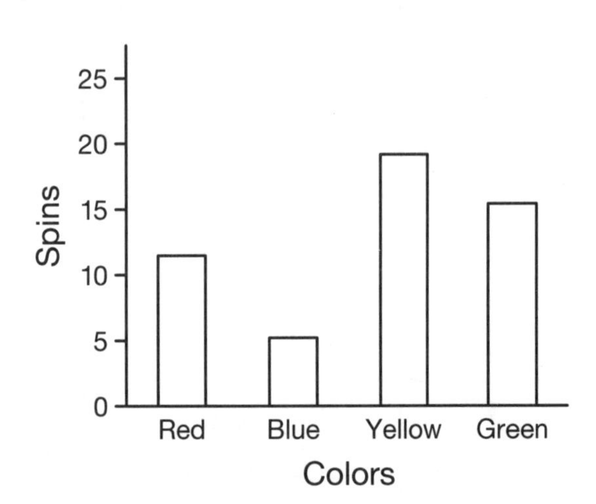


B. Use the graph to answer these questions.

1. About how many spins did Rusty and Ryan make in all?

2. Which color did they spin about twice as many times as blue?

3. Predict what color Rusty and Ryan will spin next. Give reasons for your prediction.

Explore Some More

Spin a spinner like the one Rusty and Ryan used. Spin it 20 times. Use tally marks to record your results. Compare your results with Rusty and Ryan's results. How are they alike? How are they different?

Name ____________________

Use a pencil.
Record your work.

A. Michael and Maria are rolling a die. They are using tally marks to record each number rolled. The results of each trial are shown below.

| Number on die | Times rolled |
|---|---|
| 1 | 𝍸 I |
| 2 | 𝍸 𝍸 III |
| 3 | 𝍸 IIII |
| 4 | 𝍸 𝍸 𝍸 II |
| 5 | 𝍸 𝍸 III |
| 6 | 𝍸 𝍸 |

Use the information in the table to make a bar graph. Use the data to decide on the scale to use. Label the graph and give it a title.

B. Write two questions that can be answered by looking at the bar graph. See if a partner can answer your questions.

1. ______________________________

2. ______________________________

Name ..............................................................................................

Use LacerLinks, crayons or markers, graph paper, and a pencil. Record your work.

A. Jerome, Nancy, Mallory, and Keisha made a table to show how many books they have read this month.

| Student | Number of Books Read |
|---|---|
| Jerome | 10 books |
| Nancy | 5 books |
| Mallory | 9 books |
| Keisha | 11 books |

Make a pictograph to show how many books each student has read.

1. Use cubes of one color to show how many books Jerome read. Stack the cubes.
2. Do the same for each of the other students.
3. Put the stacks side by side. Label each stack of cubes.

B. Look at the pictograph you have constructed. Write three questions that can be answered by looking at the pictograph.

1. ______________________________________________

2. ______________________________________________

3. ______________________________________________

See if a partner can answer the questions.

C. Use data from your pictograph to make a bar graph on graph paper.

Think and Write

How are your pictograph and bar graph alike? How are they different?

 

## MATH SKILLS AND UNDERSTANDINGS

- Explore how to interpret data by constructing a circle graph using LacerLinks
- Explore how to interpret data by reading and comparing information on a circle graph
- Explore combinations of a problem using LacerLinks to display data

## GETTING READY

Each student will need an index card, along with yarn and construction paper in three colors. Label three cards: **triangle, circle,** and **square.**

For the explorations, each pair of students will need LacerLinks, yarn, scissors, and crayons or markers.

For the extensions, obtain a copy of *Socrates and the Three Little Pigs* by Mitsumasa Anno. Prepare houses like the ones shown in the book and tape them to the chalkboard.

## INTRODUCING THE EXPLORATIONS

Display the cards labeled *triangle, circle,* and *square.* Ask the students to choose one of the shapes and draw it on their index cards. Categorize the cards according to the shape each student chose and organize them on the floor to make a bar graph. Discuss the information students can obtain from the data in the graph: the number of each shape, comparisons between numbers of shapes, and so on.

Next, have students sort themselves according to the shapes they drew to make a human circle graph. Use yarn on the floor to separate each section of the circle graph. Then have students fill in the yarn sections with different colors of construction paper.

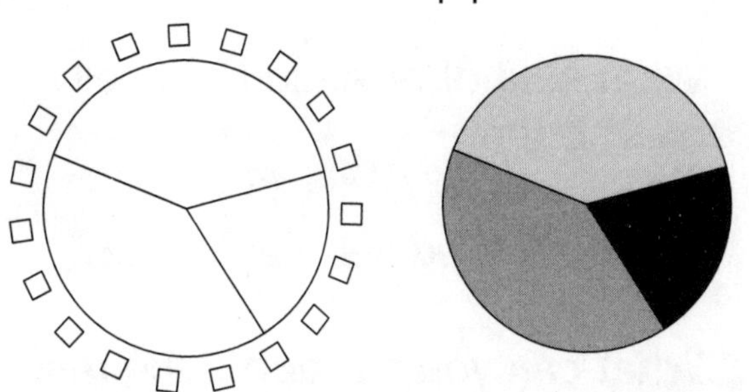

What can students tell from looking at the circle graph? Can they tell how many students drew a square? Which shape was drawn the most? Which graph shows exact numbers?

## TALKING AND WRITING ABOUT THE EXPLORATIONS

To help the students reflect upon what they have discovered during the explorations, you can ask questions for discussion or journal writing: **How are bar graphs and circle graphs alike? How are they different? When you list different combinations of things, how does making an organized list help you?**

## EXTENSION IDEAS

Read *Socrates and the Three Little Pigs* to the students. Ask the students to predict how many ways the wolf can visit the five houses. Distribute five different colored LacerLinks to pairs of students. Encourage the students to use the LacerLinks to help them keep track of the number of ways the wolf could visit the five houses. Elicit that making a list is the best way to find the number of ways the wolf could visit the five houses.

EXPLORATION 26

Name ..........

Use a pencil.
Record your work.

A. Jessica went to Funtastic Fred's Bike Store. She wanted to buy a bike horn. At the front of the store she noticed this circle graph.

What kind of bike did Fred sell most often?

What did Fred sell least often?

What can you say about the men's and women's bikes that Fred sold?

Funtastic Fred's Bike Sales

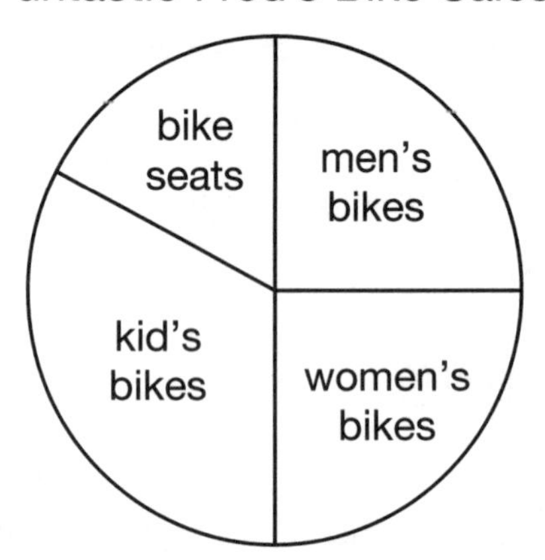


B. Jessica thought that the circle graph was a great way to show this information. She wanted Aunt Camila to display a circle graph in her candy store. Aunt Camila gave Jessica the information listed below. Read the information and see if you can figure out how to set up the circle graph to show the kinds of candy that are sold. Record your choices on the blank circle graph.

Information About Sales

- I sell more hard candy than bubble gum.
- I sell less licorice than bubble gum.
- I sell twice as much chocolate as hard candy.

Camila's Candy Corner

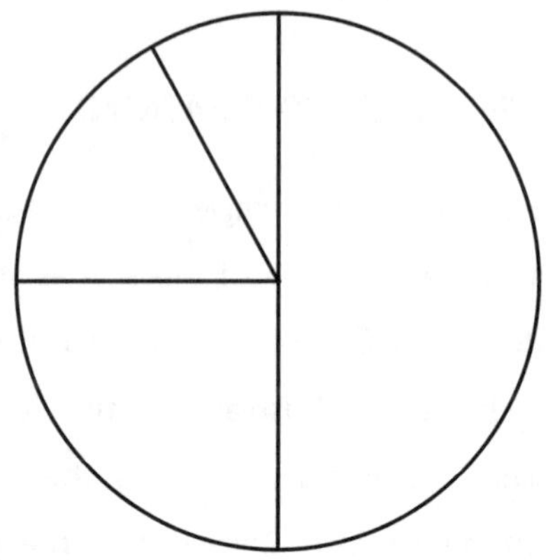

Think and Write

Why do you think a circle graph is a good way to display information?

Name ..........

Use LacerLinks, crayons or markers, circle graph paper, a ruler, and a pencil. Record your work.

A. Sara and Martin surveyed their class to find out what color they liked best. There are 16 students in the class. Here are the results.

Blue—8 Pink—4 Yellow—2 Purple—2

Make a circle graph to show the results.

1. Take one blue cube for each student who chose blue. Do the same with the other colors.
2. Put the cubes around the outside of the circle. Group them by color. Make sure the cubes are spaced evenly around the circle, like this:

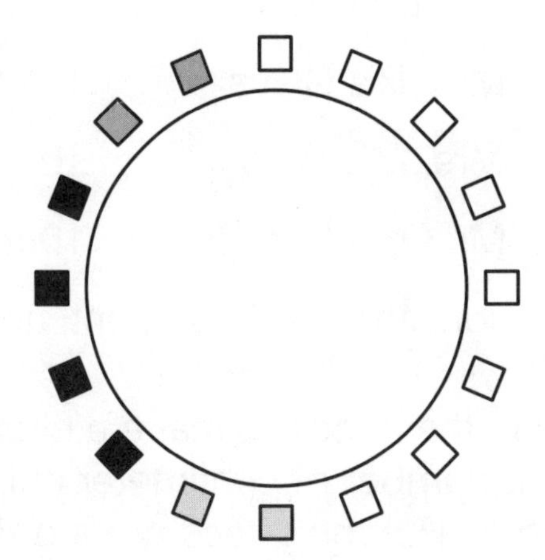

3. Place a ruler between two colors. Draw a line from the outside of the circle to the center of the circle. Do the same between the other colors.
4. Color in the sections to match the cubes on the outside of the circle. You made a circle graph!

B. What section of your graph is the largest? ______________

What two sections are about the same size as another section?

______________________________

Use fractions to describe the sections. ______________

Think and Write

Why is the circle graph a good way to display this information?

Name

Use LacerLinks, centimeter graph paper, circle graph paper, crayons or markers, and a pencil. Record your work.

Liz and Jill want to display some information about how many miles the fifth grade classes have walked in the walkathon. Liz wanted to use cubes to build a graph. Jill wanted to make a circle graph. They can't decide which graph would be the best way to display the information. They decide to make both kinds and compare them.

A. Use cubes to make a graph using the table below.

| | |
|---|---|
| Ms. L's class | 5 miles |
| Mrs. R's Class | 15 miles |
| Mrs. M's Class | 10 miles |
| Mr. O's class | 10 miles |

Record the graph on centimeter graph paper.

B. Use the cubes to make a circle graph. Gather the number of centimeter cubes for each color. Put the cubes around the outside of the circle. Group them by color. Make sure the cubes are spaced evenly around the circle.

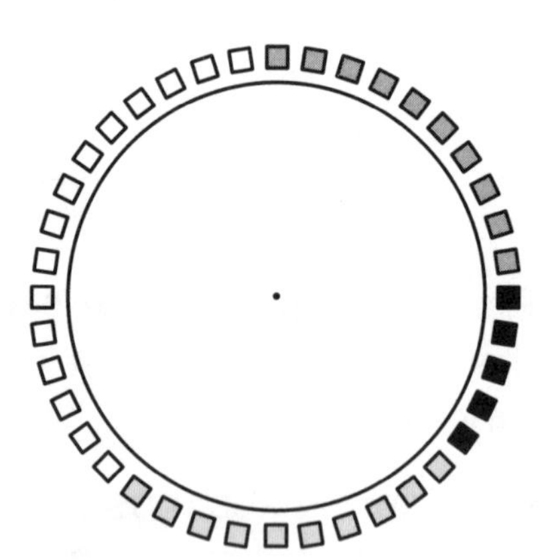

Find the center of the circle. Place a ruler between two colors and draw a line from the edge of the circle to the center of the circle. Repeat between each group of colors. Color in the sections to match the cubes on the outside of the circle.

Think and Write

Which graph shows the data best? Explain your answer.

Name ____________________

Use LacerLinks and a pencil.
Record your work.

A. Jose, Jamie, and Kim are trying to see how many different ways they can stand in a line. They decide to use cubes to help them figure out the different ways. How many different ways do you think they can stand in a line?

Write your prediction. ____________________

Put three cubes in order. Record the colors. Then put the cubes in a different order. Repeat until you have found each way.

| g | b | r |
|---|---|---|
| g |  |  |

How many different ways can Jose, Jamie, and Kim stand in line?

____________________

How did the cubes help you solve the problem?

B. Suppose Lee joins the group. How many different ways can the four of them stand in line? Use the cubes to find out. Record your work.

C. How many different ways can four people stand in line?

Think and Write

How many different numbers can you make using the digits 3, 6, and 8? Share your numbers with a partner.

Name ..........................................................................................

Use LacerLinks and a pencil.
Record your work.

A. Jenna has green, red, and yellow shorts. She also has pink, blue, and purple T-shirts. During the first two weeks of school, Jenna doesn't want to wear the same outfit twice. Will she have to?

Use cubes to help you keep track of Jenna's outfits.

| Example: | shorts | shirts |
|---|---|---|
| Monday | G | Pi |
| Tuesday | G | B |

How many outfits does Jenna have?

B. Jenna's mother buys her orange shorts and an orange T-shirt. How many outfits does Jenna have now?

Think and Write

How does making a list help you solve this problem?

## MATH SKILLS AND UNDERSTANDINGS

- Explore the concepts of central tendency by using LacerLinks to find the average, mode, and range
- Explore the concepts of central tendency by using yarn to find the average

## GETTING READY

Write statements such as the following on sentence strips:

The average temperature for June was 85 degrees.

The average amount of time you have to wait to buy a movie ticket is 4 minutes.

The average height of an eight-year-old girl is 4 feet 2 inches.

An average American laughs 15 times per day.

For the explorations each pair of students will need LacerLinks, yarn, a centimeter ruler, scissors, and crayons or markers.

## INTRODUCING THE EXPLORATION

Ask the students if they have ever heard the word *average*. Display the first sentence strip. Ask, **What does this statement say about the temperature in June? Were there any days when the temperature was different than 85 degrees?** Display the other statements one at a time and discuss what each statement means. Elicit that average stands for a group of information. For example, If the average height of an eight-year-old girl is 4'2," some girls will be taller and some will be shorter, but the average for all eight-year-old girls is 4'2."

Present the following questions and have students brainstorm the answers, **How long is a kitten? How many phone calls do you get in a day? How long does it take you to get from home to school?** List students' responses for each question on the chalkboard. Then help students find the average for each set of responses.

## TALKING AND WRITING ABOUT THE EXPLORATIONS

To help the students reflect upon what they have discovered during the exploration, you can ask questions for discussion or journal writing: **How do you find the average of something? What does average mean? What does the mode tell you about the data? Suppose the range of data is 5, what does that tell you about the data?**

## EXTENSION IDEAS

Have the students conduct a survey about the number of hours students in their class watch television in a day. Have each student use the LacerLinks to show the number of hours they watch television and find the range, median, and mode.

EXPLORATION **31**

Name ............................................................

Use LacerLinks and a pencil.
Record your work.

A. Samantha, Alexis, Jamal, and Henry have been playing a game. They roll a die. Each player takes the number of cubes to match the number that he or she rolled. The first person to get 20 cubes wins. Here is what each player has so far.

Samantha—15  Alexis—9  Jamal—12  Henry—12

Use cubes to make a stack for each person.
What do you notice about the stacks?

Which two people have the same number of cubes?

When two of the categories have the same number, this number is called the **mode.**

Looking at the game so far, who do you think will win?

B. Jane and Elisa decide to play the game with Samantha, Alexis, Jamal, and Henry. Here is what each player has after 3 rolls.

Samantha—10  Alexis—7  Jamal—14  Henry—7
Jane—11  Elisa—15

Record what each player has on a line plot.

Samantha  Alexis  Jamal  Henry  Jane  Elisa

What is the mode?

Think and Write

Looking at the game so far, who do you think will win? Why?

Name

Use LacerLinks, crayons or markers, centimeter graph paper, and a pencil. Record your work.

A. Carlos has 4 stacks of cubes.

black orange yellow red

Carlos wants to give each of his friends a stack of cubes. He wants to have the same number of cubes in each stack.

Help Carlos. Make stacks of cubes to match his stacks.

Now think of a way to rearrange the cubes to make the four stacks even. Describe or draw your idea. Then try it.

How many cubes are in each stack now? _____ The number of cubes in each stack is the average number of cubes in the four stacks.

B. Find the average number of cubes in the four stacks described below. First make the stacks. Then rearrange the cubes to make the stacks even.

- Stack A has 5 cubes.
- Stack B has 2 cubes.
- Stack C has 3 cubes.
- Stack D has 2 cubes.

What is the average number of cubes in the four stacks?

Think and Write

How did you move the cubes to find the average number of cubes in the stacks?

# EXPLORATION 33

Name ........................................................................

Use yarn, a centimeter ruler, and a pencil.
Record your work.

Tina made friendship bracelets for three of her friends. She wanted each bracelet to fit her friends' wrists perfectly, so she carefully measured their wrists. Her bracelets were a big hit. She decided to make more. To make it easier, she decided to find the average size of her friends' wrists. Help Tina find the average size.

1. Find three friends.

2. Use yarn to measure around each person's wrist. Don't forget to measure your own wrist! Cut the yarn to fit each person's wrist.
3. Lay the yarn pieces end-to-end.

4. Measure a new piece of yarn that is the same length as the yarn pieces.

5. Fold the yarn in half. Then fold the yarn in half again.

6. The folded length of the yarn is the average wrist size.
7. Use a ruler to measure the folded yarn piece to find the length in centimeters.

_________ cm

Think and Write

How does the yarn help you find the average size?

EXPLORATION 34

Name ..............................................................

Use LacerLinks and a pencil.
Record your work.

A. Ms. Suarez's students are estimating the number of cubes they can hold in their hands. Here are the estimates for a group of four students.

10 12 11 15

Find the average estimate for the number of cubes that the four students can hold in their hands. Add up all of the estimates. Then divide by the total number of estimates.

Here are the exact numbers of cubes that the students can hold in their hands.

8 10 9 13

What is the average number of cubes that the students can hold in their hands? Add and then divide.

B. Work in a group of four. Record each person's estimate for the number of cubes he or she can hold. Find the average estimate for your group of four.

Now each take a handful of cubes. How many can each of you hold? Record the exact number of cubes for each person.

What is the average number that each person can hold?

How does the estimate compare with the average number of cubes that each person can hold? Is the estimate more or less than the exact amount?

Explore Some More

Find some other kind of cubes or buttons to use. How many do you think you can hold? How does knowing how many cubes you can hold help you make your estimate? Estimate the number and figure out the average. Now take a handful of the things you are using. Record how many you can hold. Find the average for the actual amount. How do the averages for the estimates and actual amounts compare?

Name

Use LacerLinks and a pencil.
Record your work.

A. Mr. Ortega put some cubes in a jar. The students estimated how many cubes were in the jar. Here are their estimates.

| 24 | 32 | 18 | 28 | 29 | 42 |
|---|---|---|---|---|---|
| 34 | 44 | 34 | 43 | 25 | 31 |

Make a line plot to help you organize the data.

15 20 25 30 35 40 45 50

Look at the line plot.
What is the largest estimate? ________

What is the smallest estimate? ________
To find the range of estimates, subtract the smallest estimate from the largest estimate. What is the range for this data?

How did making a line plot help you?

B. Each student in Mr. Ortega's class got a bag of cubes. The students estimated the number of cubes. Then they counted the exact number. The table below shows the results for 10 pairs of students.

| **Estimate** | 23 | 34 | 25 | 34 | 21 | 35 | 23 | 25 | 18 | 35 |
|---|---|---|---|---|---|---|---|---|---|---|
| **Exact** | 25 | 24 | 26 | 27 | 23 | 33 | 19 | 20 | 20 | 30 |

Find the range for the estimates.
Find the range for the exact amounts.
Which had a larger range?

Think and Write

What does the range tell you about a set of data? If you know that the range for a group of estimates is 5, what does that tell you? If the range for a group of estimates is 35, what does that tell you?

 

# NOTES ABOUT EXPLORATIONS **36-40**

## MATH SKILLS AND UNDERSTANDINGS

- Explore the concepts of displaying data by constructing a line graph using data from a table
- Explore the concepts of displaying data by comparing a bar graph and a circle graph
- Explore the concepts of displaying data by choosing the best graph for a given situation

## GETTING READY

List the months on the chalkboard. Record the number of students having a birthday in each month. Use tally marks or numerals to record the data.

For the explorations, each pair of students will need blank paper and crayons or markers.

## INTRODUCING THE EXPLORATIONS

Divide the class into groups of four. Assign each group a type of graph: bar graph, pictograph, circle graph, and line plot. Have each group use the birthday data to make the assigned graph. Remind the students who are making the circle graph to use cubes.

Display the graphs. Ask, **Which graph shows the exact number of birthdays in each month? Which graph is easy to read? Does one graph show the data better than another graph? Why or why not?**

## TALKING AND WRITING ABOUT THE EXPLORATIONS

To help the students reflect upon what they have discovered during the explorations, you can ask questions for discussion or journal writing: **How are a circle and bar graph alike? How are they different? Is it easier to make a bar graph or a circle graph? If you were to choose a graph to show the heights of the students, what type of graph would you choose? Give reasons for your choice.**

## EXTENSION IDEAS

Ask the students to look in newspapers and magazines for different types of graphs. Have each student bring in an example. In groups, have students discuss how well the graph displays the data. Would they choose another type of graph? If so, why? Have each group share their ideas with the class.

Name ..............................

Use a ruler and a pencil.
Record your work.

A. Every day at 9:00 a.m., Alfie and Tiffany find out the temperature. They use a table to help them keep track.

| Day | Mon. | Tues. | Wed. | Thurs. | Fri. |
|---|---|---|---|---|---|
| **Temperature** | 32 | 38 | 34 | 28 | 32 |

Use the data from the table to make a line graph. Label the side and bottom of the graph. Plot the temperature for each day. Then use a ruler to connect the points. Add a title to the graph.

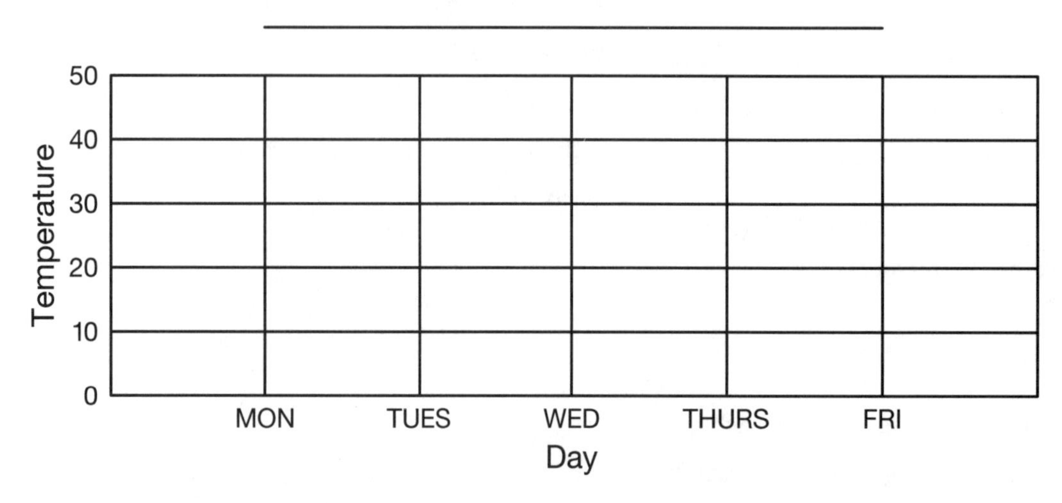


B. How did you decide where to put 38 degrees on the graph?

What is the range of temperatures for the week?

What temperature occurred more than once?

What is the difference between the temperature on Tuesday and the temperature on Wednesday?

Think and Write

Write two questions that can be answered by looking at this graph. See if a partner can answer your questions.

 

EXPLORATION **37**

Name ............................................

Use a pencil.
Record your work.

A. This graph was on the sports page of the newspaper. Read the three descriptions below to decide which description matches best.

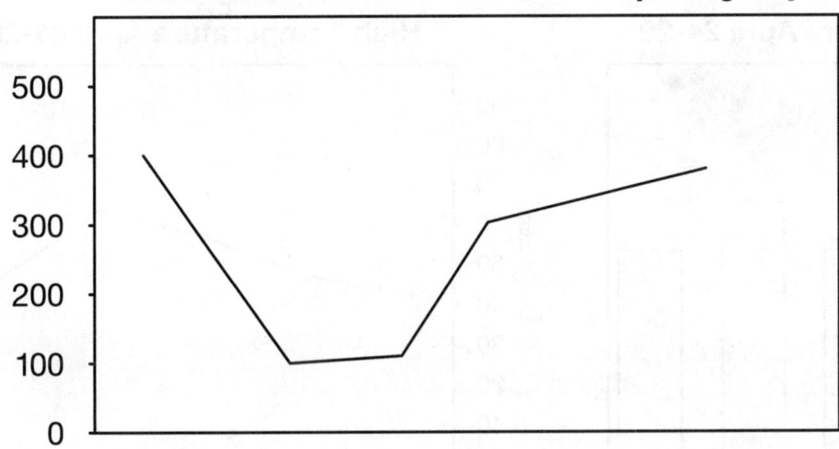


1. Attendance was high the first year but fell the last year.
2. Attendance was high the first year, fell a little bit, and then dropped to an all-time low.
3. Attendance was high the first year, fell for a few years, and then went up again.

Which description best describes the line graph? Why do you think so?

B. Look at the line graph below. Write a description to match the data.

Description:

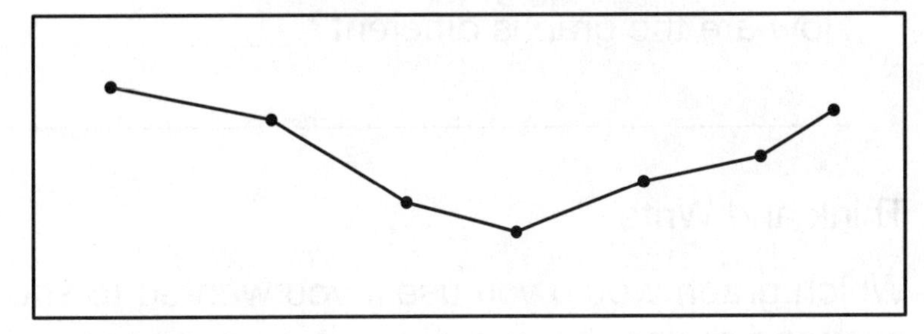


Explore Some More

Make a line graph. Write one sentence that describes your graph and two that do not. See if a partner can match your graph to the sentence that describes it.

Name ..........................................

Use a pencil.
Record your work.

A. You can use graphs to represent data in different ways. The graphs below show the same data.

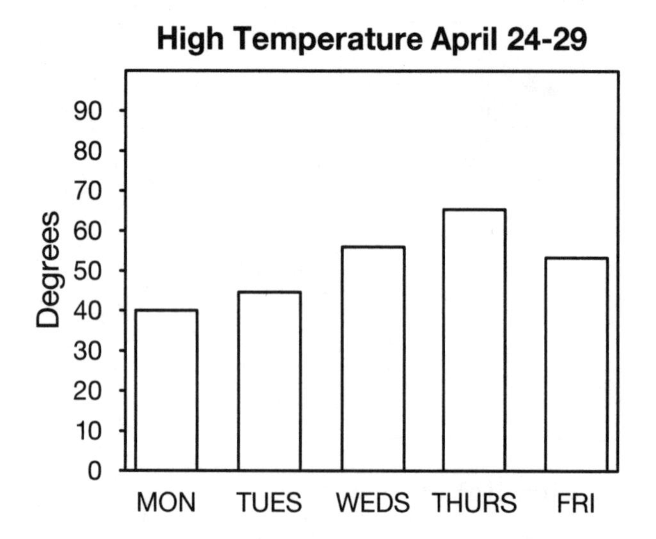


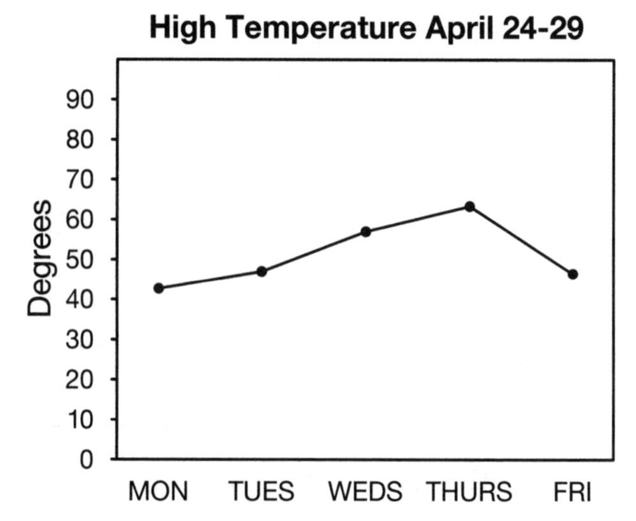


What information does each graph tell you?

B. How are the graphs alike? ____________________

____________________

How are the graphs different? ____________________

____________________

Think and Write

Which graph would you use if you wanted to show how the temperature changed during the week? Explain your answer.

EXPLORATION 39

Name ______________________________

Use a pencil.
Record your work.

A. Bar graphs compare data. Line graphs show how data increases or decreases over time. Circle graphs show parts of a whole. Below are some graphs. Study each one.

**Graph A**

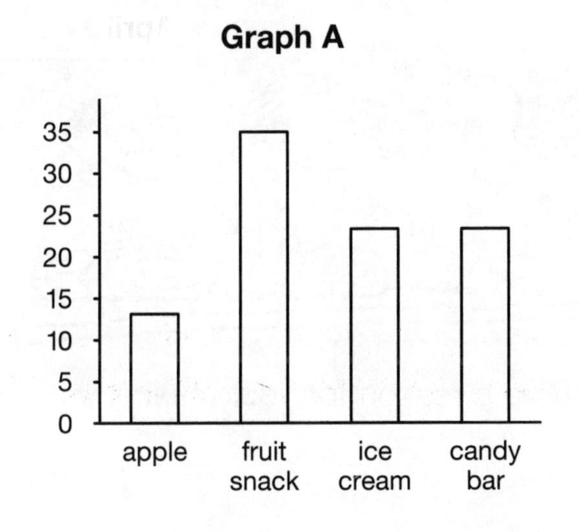


**Graph B**

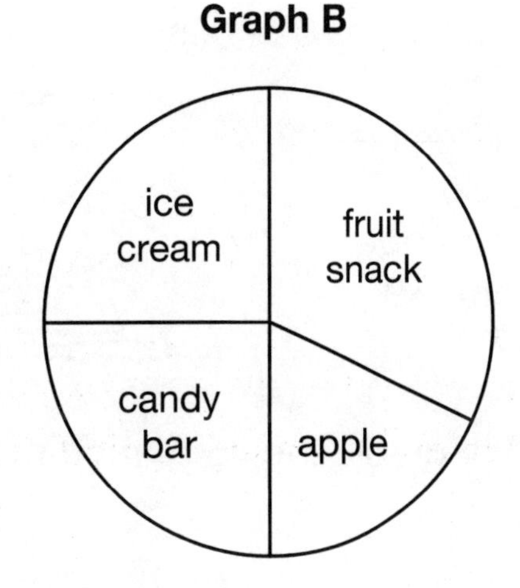


Match each graph to the sentence that best describes it. Watch out! One sentence does not have a graph to match it.

1. This graph shows how the sale of apples changed from month to month.
2. This graph compares four different snacks sold at the snack bar.
3. This graph shows all of the snacks sold and shows what part each snack is of the whole.

B. Read the sentence that does not have a graph to match it. Make a graph to fit the sentence.

Think and Write

Write a question that can be answered by your graph. See if a partner can use your graph to answer the question.

Name ............................................................

Use a pencil.
Record your work.

A. Ahmad noticed that more children buy the school lunch on days when either pizza or chicken nuggets are served. Pizza and chicken nuggets are served only once a month. Ahmad wants the cook to make them more often. He has collected data to show the school lunches that the children in his class like best.

Grilled cheese—12
Pizza—30
Chicken nuggets—27
Spaghetti—9
Chili—7

What kind of graph should Ahmad make? Give a reason for your choice.

B. Make the graph you chose. Write three questions that can be answered by your graph.

Explore Some More

Compare your graph to a friend's. Did you make the same graph or a different one? Which graph best shows the data? Why?

 

POPCORN GLYPH

- **Description:** This investigation centers around *The Popcorn Book* by Tomie de Paola. The book presents information about the origin of popcorn, in what region the most popcorn is eaten, and how popcorn is prepared. After the students read or listen to *The Popcorn Book,* they examine individual bags of popcorn containing at least three different brands. The students estimate the number of popcorn pieces in each bag. They then determine the number of cups of popped corn in the bag, calories per serving, cost, and type of popcorn. Using this information, the students construct a glyph for each brand of popcorn. Finally the students compare the popcorn glyphs and draw conclusions about the three different brands of popcorn. The investigation can end with a taste test, and students can show their preferences in a bar graph.
- **Skills Developed:** The students estimate quantities and develop estimation strategies, collect data and represent their information in a glyph, interpret glyphs and use them to make comparisons; they may also construct a line plot and bar graph.
- **Getting Ready:** Have the students work in groups of four. Each group of students will need a bag of popped popcorn. Bags of popped popcorn or microwave popcorn are the easiest for the students to use because all of the necessary information is printed on the bag. At least three different brands should be used. Also have available crayons or markers and paper on which students can draw a popcorn glyph and key or a copy that you have prepared.

### Finding Out About Popcorn

Ask the students if they like popcorn. Have them brainstorm everything they know about popcorn and record their ideas in a web.

Read aloud *The Popcorn Book* or have students read it.

Discuss the information with the students. Ask, **What new information did you learn about popcorn from this book? Did any of the information surprise you?**

Give a closed bag of popped popcorn to the students. Say, **In your group, estimate how many pieces of popcorn are in the bag without opening it up.** After the students have made their estimates, ask how they determined their answer. Say, **Now, take out one handful of popcorn and count the pieces. Make a new estimate.**

Have the groups discuss their strategies for estimating. Then have them count the pieces of popcorn in the bag. Discuss with the students those estimates that were reasonable and those strategies that were most helpful. Encourage the students to make a bar graph that shows their estimates and the actual number of pieces.

**Popcorn Glyph Key**

**Calories:** one C for every 10 calories
**Fat Content:** one stripe down the popcorn box for each gram of fat
**Price per package:** one $ along the bottom of the box for each $.50 spent
**Butter/No Butter:** butter—color the popcorn yellow
**No Butter**—leave it plain

Have the groups brainstorm the information that can be found on the package (calories, fat, ingredients, number of cups popped, and so on). Say, **Make a popcorn glyph to compare the different brands of popcorn.** Display the popcorn glyph key and popcorn box. Discuss how the key is used to make the glyph. Then have each group make a glyph that represents their brand of popcorn.

When the groups have completed their glyphs, display them where everyone can see them. Compare the completed glyphs. Ask, **What can you tell about the different brands of popcorn? Which ones have the most popped pieces? Which brand has the least amount of fat? Which brand is most expensive?** Have the students discuss how the glyphs help them to compare a great deal of information in an easy way.

### Conduct a Taste Test

Students can taste the different brands of popcorn and rate their preferences. Have students tally their results and use the data to make a line plot or bar graph. Students might also want to design their own glyph key and glyph.

AVERAGING ANYONE?

- **Description:** This investigation centers around the book *Averages* by Jane Jonas Srivastava, which presents many examples of averages. After the students have read or listened to *Averages,* they will discuss what the term average means. Small groups will brainstorm averages they could calculate using data collected in class: height, hand span, shoe size, long jump length, heart rate, number of blinks per minute, number of books read in a month, number of times moved in a lifetime. After choosing a category to explore, students will collect the data and find the average. Finally, they will present the information on a page that can be used to make a classroom Big Book of Averages.

- **Skills Developed:** Students determine average from a set of data they collect.

- **Getting Ready:** Read *Averages.* Discuss the information with the students. Ask, **What new information did you learn from this book? Did any of the information surprise you? What does it mean when we say the word *average*?**

### Making a Big Book of Averages

Divide the class into small groups and have students think of a category they can use to determine an average for the class.

Have the groups determine how they will collect the data. Have the groups list their categories on the chalk-

board to ensure that there are no repeated questions and that the students' ideas are reasonable. As a group, decide how the information will be collected. For example, if the students need to answer a question—"How many times have you moved since you were born?" or "How many hours of television do you watch per day?"— students could take a survey. If data such as "How tall are you?" or "How many times can you blink in a minute?" needs to be collected, data stations could be set up. After all the data has been collected, ask, **Now that each group has the data that has been collected for its category how can you find the average? Is there a special tool that might help you find the average?** Distribute calculators to the student groups and model for the students how to find the average by adding up the data and dividing by the number of pieces of data. Allow the student groups to find the average. Distribute a large piece of paper to the students. Say, **This piece of paper will be one page in our classroom Big Book of Averages. Think of a meaningful way to display your information.**

When each group has completed its page, have students share their information. Ask, **Are you surprised by the information? What new information did you learn? Do you think the average for our class will be the same for another class? Why or why not?**

Finally, put all of the student pages together to create a book. Students might want to check the book out for one night to share and read with their parents.

# INVESTIGATION 3

DESIGN A GAME

- **Description:** This investigation explores games and the concept of chance, probability, and fairness in setting up directions. The students will play different games in groups of four and discuss the characteristics of each one. Specifically, chance will be discussed and the students will investigate the role of dice, spinners, or drawing cards in each game. The students will design and construct a game in their groups, making sure the game is fair. Finally, the students will play the games designed by each group.
- **Skills Developed:** The students determine the characteristic of a game, design a game that uses dice or spinners, and write the directions for play. Students also examine games and determine its fairness.
- **Getting Ready:** Have the students work in groups of four. Distribute a different game to each group. Each game should have a spinner, dice, or cards that are drawn at random. Have the students play the games so that everyone gets a chance to participate. Ask, **Is the game you are playing fair? Are all of the players equally likely to win? How do you know?**

### Making a Game

Have each group generate a list that begins, *A game must have* . . . . When the groups have completed the lists, record their responses. Responses should include an equal chance for everyone to win, a clear object, directions for play, and so on.

Tell the students they are going to design and construct a game. They must make sure they specify all of the parts of the game, such as spinners, dice, and so on. Distribute pieces of cardboard, construction paper, glue, and scissors for the students to use. Allow ample time for the students to design and construct the game.

Have each group demonstrate its completed game to the class. Then have the groups rotate so that they can play each game.

After all of the students have played all of the games, refer to the list generated by the student groups to discuss each one. Ask, **Do the games contain all of the necessary parts? Are the games fair? Why or why not? Does every player have an equal chance of winning?**

### Game Day

Invite another class to come and play the games.

TAKE A CENSUS

- **Description:** This investigation asks the students to develop and conduct a census for their classroom and two other classrooms, preferably of different grade levels. Students will discuss what a census is and what its purpose is. They will discuss how a survey is used to gather information. Students will write questions for a census survey, conduct the survey, and display the information in a graph. Finally, the students will compare the census information from one classroom to another.
- **Skills Developed:** The students will develop and conduct a survey. The students will conduct the survey, tally the results, and display the information in a graph. They will compare the results for each classroom.
- **Getting Ready:** Ask the students if they have ever heard of the census. Tell them that a census is used to find out how many people live in the United States, what kinds of jobs they have, where they live, and so on. Ask, **How do you think census information is collected?** Tell the students that the information is collected through a survey and, in some cases, through a personal interview. Tell the students that they are going to conduct a census in their classroom and in at least one other classroom. Ask, **Do you think the census data we collect will be the same or different from class to class?**

## Collecting and Sharing the Data

Divide the class into groups of four. Have the students brainstorm a list of questions for their census and share their lists with the class. As a class, decide what questions will be asked in the census. Prepare a final list of census questions and have student groups conduct the survey in their own classroom and in at least one other classroom in the school.

When surveys are complete, model how to tabulate the results of one question. Then have students work together to tabulate and record the survey results, keeping the surveys for each class separate. When all of the results have been recorded, ask, **Is there a certain kind of graph that would best display your data so that you can show how our class's data compares to another class's data?** Have the student groups decide on a graph and construct a graph to display the data.

## Present the Results

Share the results of the census by publishing a newsletter that describes and displays the results.

 